# THE AUTO EMISSIONS BIBLE

## How to Pass the Vehicle Emissions Test

*Sam Bell*
*Ralph Birnbaum*

# The Auto Emissions Bible

ISBN-13: 978-1468130188

ISBN-10: 1468130188

# About the Authors

**Sam Bell** has owned and operated The Lusty Wrench, a premium auto repair facility in Cleveland Heights, Ohio for 32 years. In addition to the normal ebb and flow of common maintenance and repair tickets, Sam is often called upon to diagnose and repair driveability issues that stump other local repair shops.

**Awards and Experience:** Sam was named the 2010 Delmar/Cengage ASE Technician of the Year for earning the highest composite score on the ASE certification exams. Sam had another perfect score on the L1 (Advanced Engine Performance Specialist) test, and his overall scores were the highest of more than 350,000 professionals who took the tests.

**Publishing Credentials:** Sam is a frequent contributor to *Motor* magazine, and has garnered five International Automotive Media Awards, including a Gold Medallion. He has also won three ASBPE Awards (American Association of Business Publications Editors), including a Gold Medal. He teaches Ohio One training courses on driveability and emissions.

**Ralph Birnbaum** has 40 years of automotive experience, and that includes real experience fixing cars as a master automotive technician, specializing in auto electric and vehicle performance issues. His is a former tech editor and editor of a national trade publication, and has been researching and preparing books and training materials for two decades.

He is currently Automotive Editor at ECS Tuning, where he prepares ad copy and training materials. He also contributes technical illustrations used in Standard Motor Products automotive training classes. Ralph still has all his tools, and regularly gets his hands dirty fixing real car problems.

**Publishing Credentials:** Writing credits include numerous automotive articles and over 20 training programs for professional technicians. His training materials have been used in both corporate and state emissions training programs.

# TABLE OF CONTENTS

### Overview or Just-in-Time Reference—You Decide
Too much to read; too little time? Use this book one of two ways:

- **If you're an auto repair novice,** use this as a course book: start at the beginning and read straight through.

- **If you already have a working knowledge of vehicle repair basics,** need to refresh your memory, or have a specific question, use this book as a just-in-time quick reference.

Need an advanced search? Individual topics and subtopics are listed in the index by key word.

### Special Symbols
You'll see special symbols in the margins. Here's what they mean:

This symbol indicates a very tasty piece of information that deserves special attention.

We'll place the **NEW** symbol next to any information that represents a recent change in standards or technology. OBD II is a work in progress: expect the OBD standards and their application to evolve and grow.

### Glossary
See an unfamiliar term? Check the Glossary starting on page 281 for a list of common terms and definitions.

### Gray Boxes!
Gray boxes contain OBD II Diagnostic Trouble Codes that are commonly associated with the condition or fault being described on a page. These are not the only possible codes, but they are likely suspects!

These DTCs are commonly associated with secondary AIR failures:

**P0410-P0419 -** Miscellaneous AIR failures

## Safety First

Some folks scoff at the safety warnings at the beginning of auto repair books. They consider them to be an annoying waste of time, paper, and ink. As professional auto technicians with multiple scars and assorted aches and pains, we can assure you that your body is not indestructible.

No one cares more about your safety that you do. Nobody. Get into the habit of wearing good, wraparound safety goggles. Wear gloves and skin protection. Never work alone. Work in a clean, well-ventilated area with good illumination. Wipe up fluid spills immediately. Have a shop-approved fire extinguisher and eye wash station handy at all times.

Safety cautions apply to professional auto repair technicians and prosumers alike.

## Zapp!

When it comes to safety, we need to add a special caution about high voltage systems in electric and hybrid-electric vehicles. Cars like the Prius, Insight, Fusion Hybrid, and other Hybrid Electric Vehicles (HEV) have *several hundred volts* available in their batteries, and potentially higher voltages in their main cables, and motor/generators.

We strongly recommend that these systems be serviced only by trained, properly equipped service personnel. That doesn't mean you can't go out and get the training and equipment to become a hybrid specialist. But the dangers posed to the human body by 150-400+ plus volts are real, and not to be taken lightly. We're talking about serious injury or possible death from electrocution here, so please take our warnings seriously.

## Fuel System Cautions

Exercise caution when working around pressurized fuel systems. Recent pressure increases in both diesel and gasoline engine fuel systems require special gauges and test procedures. Plug the wrong gauge into a high pressure system and it may blow up in your hand. Check OEM system specifications and limit your tests to approved procedures, using appropriate equipment.

# Introduction - Preventing Problems

Before we start discussing ways to diagnose and repair emissions problems, let's look at a list of common maintenance procedures that **prevent emissions problems** in the first place!

- **EGR passage cleaning**: Particularly recommended where the EGR does not flow into the intake manifold plenum, but through distribution passages that are drilled or cast inside the manifold. Remove light carbon deposits with chemical cleaners. Heavier, hardened deposits may require rodding or scraping.

- **PCV valves**: Use only high quality PCV valves. Failed emissions tests, oil leaks, and oil consumption problems can often be traced to a faulty PCV valve. When installing a new valve, do a thorough job and clean the crankcase ventilation hoses. Replace hoses that are oil softened, heat hardened, cracked, or loose. Check and clean valve cover breather-baffle restrictions in engines that use them instead of a PCV valve. Clean the valve cover vapor passages, and blow ventilation hoses clean with compressed air.

- **Oil, air, and fuel filter changes:** Prevent problems and excessive engine wear: change fluids and filters at recommended intervals, based on OEM recommendations, fine-tuned for vehicle use and operating environment. Gasoline dilution of crankcase oil causes rich running as fuel vapors are drawn back into the intake manifold through the PCV system. Use only the correct API oil grade and recommended viscosity oil.

- **Valve lash adjustments:** Valve lash affects cylinder sealing and valve timing, both of which have a major impact on performance and emissions.

- **Spark plugs, distributor caps and rotors, plug wires:** Worn out spark plugs or damaged secondary ignition components like caps, rotors, and plug wires lie at the root of many emissions problems. Maintain the ignition system for reliability, economy, and low emissions.

- **Ignition timing:** Base ignition timing and computer timing correction in a running engine have a big effect on emissions. Base timing changes as timing belts or chains stretch. Incorrect base timing can also result from failure to observe recommended timing adjustment procedures.

- **Oxygen sensor replacement:** Oxygen sensors degrade over time. Their signal output range may decrease or shift. They will eventually lose their quick reflexes and respond sluggishly to changes in exhaust oxygen content. Heated sensors tend to last longer, but any sensor with over 60,000 miles of use should be tested with a lab scope to confirm its minimum and maximum voltages, response times, and switch rates.

- **Throttle body and top engine cleaning** help improve idle quality and throttle response. Throttle plates, throttle bodies, and intake manifolds all benefit from periodic cleaning. Removing carbon deposits also prevents throttle plate sticking and just-off-idle vehicle hesitation.

- **Carbon cleaning/injector cleaning:** Both fuel injectors and intake manifold air passages should be cleaned at regular intervals. Cleaners reduce intake valve deposits that lead to cold stalling and tip-in hesitation.

Here's a list of common tests you need to know—and use—if you intend to repair emissions problems successfully. Some are discussed in this book. If you are not yet familiar with these procedures, do some research. Get your hands on the necessary equipment, and practice the techniques associated with these tests.

They are the core skills needed to successfully diagnose and repair the root causes for vehicle emissions test failures.

How many of these valuable tools do you use on a regular basis?

- Fuel pump pressure test
- Fuel pump volume test
- Fuel pump voltage supply test
- Engine vacuum test
- Cranking compression test
- Running compression test
- Cylinder balance test
- Cylinder leak-down test
- Injector balance/injector volume test
- Ignition timing adjustment
- Ignition firing voltage, including "snap throttle KV demand" test
- Ignition coil available KV test
- Ignition system spark KV leakage test
- Circuit voltage drop test
- Exhaust backpressure test
- Alternator AC voltage ripple test
- Alternator charging voltage and current test
- Battery state of charge and load test
- Lab scope tests: injector voltage and current waveforms
- Fuel pump current ("amp ramp") test
- Ignition coil current ramping test
- Scan tool tests, including bi-directional (functional) testing
- Catalytic converter oxygen storage capacity test
- Catalytic converter cranking hydrocarbon efficiency test

# COMBUSTION 1

# Combustion

## Combustion Triangle
**Modern engines run stronger and burn cleaner than ever!**

To understand how engines release the energy stored in fuel, we'll introduce you to the **combustion triangle**. It explains the relationship among air, fuel, and heat that makes the internal combustion engine the durable work horse of our fleet.

- **Fuel** - The fuel used in most spark ignition engines is gasoline.

- **Oxygen** - To burn one gallon of gasoline, expect to consume about 9,000 gallons of air. (Now you know why air filters get dirty so quickly!)

- **Heat** - The heat required to start the combustion process comes from a tiny lightning bolt that jumps the spark plug gap. The spark ignites an atomized, pressurized mixture of air and fuel. The ignition system must be capable of producing and delivering reliable sparks of 40,000 volts, or more!

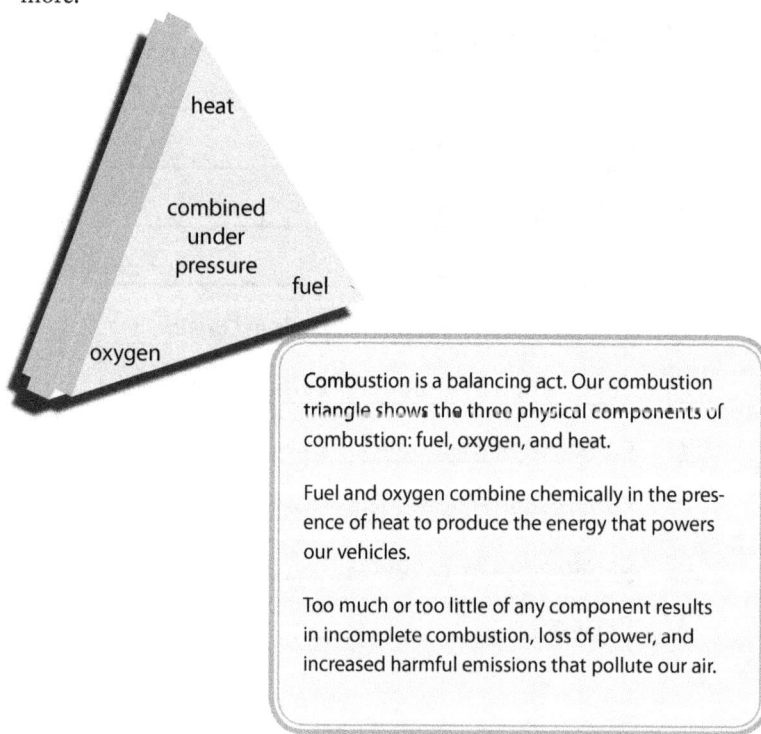

heat

combined
under
pressure

fuel

oxygen

Combustion is a balancing act. Our combustion triangle shows the three physical components of combustion: fuel, oxygen, and heat.

Fuel and oxygen combine chemically in the presence of heat to produce the energy that powers our vehicles.

Too much or too little of any component results in incomplete combustion, loss of power, and increased harmful emissions that pollute our air.

12

# Combustion

## Oxygen

The oxygen needed for combustion comes from the atmosphere that surrounds us. An internal combustion engine is an air pump. A large volume of air must be drawn into the engine to provide the oxygen for combustion.

- Most of the air, about 78% of it, consists of **nitrogen**, whose chemical symbol is N or $N_2$.
- About 20 to 21% of the atmosphere is **oxygen**. The chemical symbol for oxygen is O or $O_2$. (The subscript 2 indicates that it takes two atoms of oxygen or of nitrogen to make a single molecule of either.)
- The remaining tiny fraction of atmospheric gases is a mixture of many elements and chemical compounds.

## Fuel

Fuel is a complicated chemical "soup" made of thousands of chemical compounds known as **hydrocarbons**. Each has its own specific chemical formula. To keep things simple, we'll refer to them in a group that we'll label **HC**.

In simplest terms, HC is fuel.

**HYDROCARBONS**

Modern fuel fill hoses at your local gas station are fitted with a collar to catch fuel vapors before they can escape to the atmosphere.

# Combustion - Ignition

## Ignition

The third and final ingredient of our ignition triangle is **heat**. In spark-ignition gasoline engines, air and fuel are compressed inside the engine's cylinders and then ignited by a spark from an ignition system. The ignition system commonly includes a step-up transformer known as an **ignition coil** that is connected to a spark plug at each cylinder.

There are two critical parts to spark-ignition:

• Spark energy (the *amount* of electrical energy available in the spark).

• Spark timing (*when* the spark occurs during the combustion cycle).

Diesels are compression-ignition engines and have no external ignition system. Instead, they squeeze the air/fuel mixture inside each cylinder well beyond the compression ratios of spark-ignition gasoline engines. Increasing cylinder pressure increases cylinder heat until the mixture gets hot enough to burn. (See Chapter 14 for more on diesel emissions.)

## Compression

While not an *ingredient*, compression is a **condition** required for combustion. Low compression resulting from an engine mechanical failure reduces combustion efficiency, and may prevent combustion altogether.

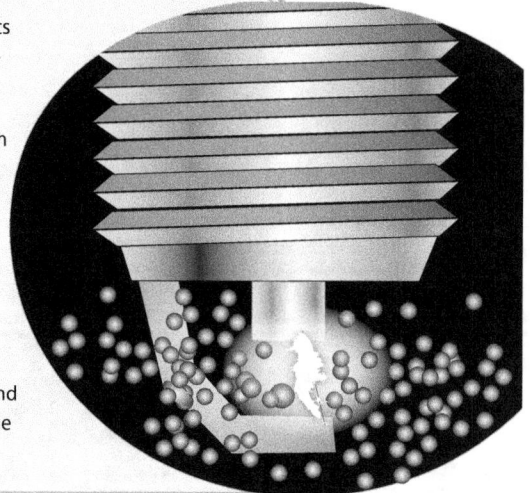

It's the zap in the gap that starts the burning process in a spark-ignition engine.

The spark must contain enough energy to cause a rapid oxidation of air and fuel.

Spark timing is a critical factor, affecting both performance and emissions.

Misfiring cylinders do not produce power, sending raw and partially combusted fuel out the exhaust.

# Combustion - Engine Cycles

## Suck-Squeeze-Bang-Blow

The internal combustion engine is still king of the hill. Internal combustion has converted fossil fuels into power for our cars and trucks for more than a century. Early engines used carburetors and magnetos or breaker point ignitions to fuel and fire combustion.

In a modern vehicle, the Powertrain Control Module (PCM) directs solenoid-operated fuel injectors and electronic ignition systems to keep fuel and fire synchronized with engine mechanical movements. The entire process is as finely tuned as the performance of a symphony orchestra.

Internal combustion hinges on mechanical parts of the engine working together as a large air pump. Our cars and trucks are powered by four stroke engines.

The four strokes are: Intake, Compression, Power, and Exhaust. The slang equivalent to this list: Suck, Squeeze, Bang, Blow.

## How It Works

- Pistons move down inside cylinder-shaped bores, ingesting atmospheric air.

- Atomized fuel is added to the air. This volatile mixture is **squeezed** and ignited — **BANG.** The resulting explosion strikes the head of a piston, forcing it toward the crankshaft. A rod connecting piston and crankshaft rotates the shaft, and it turns.

- At the end of each combustion cycle, the trash needs to be thrown out! Burned out gases are **blown** from the cylinder to the exhaust system, so the process can repeat itself.

Let's look at the four combustion strokes in more detail.

# Combustion

### Intake Stroke

The first step in combustion is to fill a cylinder with an **explosive charge** of **air** and atomized **fuel**. (If you've ever blown up a tin can with a cherry bomb, you understand the effects of an explosion in a container.)

A cylinder is filled with air when its piston moves toward the crankshaft (downward in most engines), creating a low pressure inside its cylindrical bore. As the piston moves, the cylinder's intake valve opens, allowing atmospheric pressure to **push air into the cylinder.**

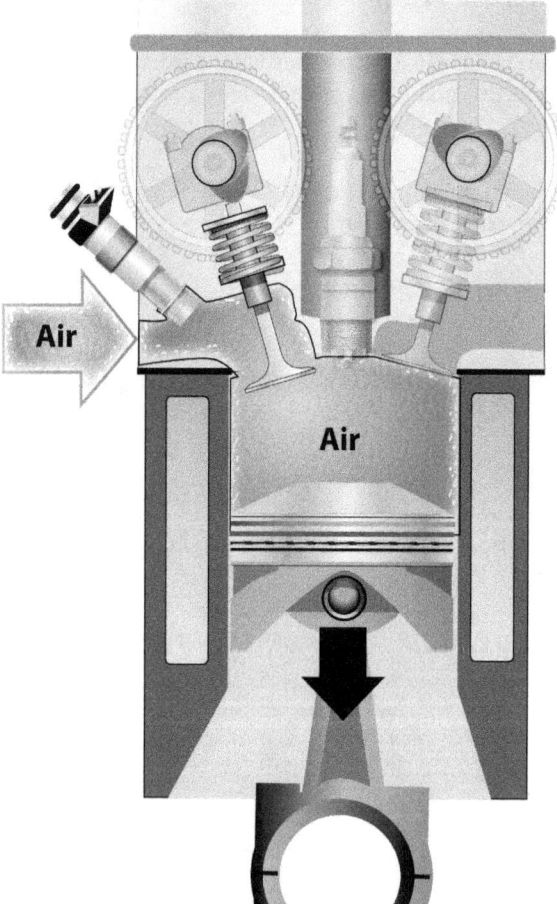

The computer briefly operates a fuel injector to spray atomized fuel into the incoming air. The injector stays open for a very short time, so short in fact, that its open time, or ON-time, is measured in *milliseconds* (thousandths of a second).

So far, so good. We now have a combustible volume of air and atomized fuel inside the cylinder.

## Compression Stroke

Now it's time to put the squeeze on the gassy mixture in the cylinder. The intake valve that opened to let the air and fuel enter the cylinder during the intake stroke now closes. With both the intake and exhaust valves closed, the cylinder is sealed.

The piston reaches the bottom of the cylinder bore and reverses direction. As it moves upward, it begins to squeeze the air and fuel trapped inside the cylinder. Compression creates heat. The greater the compression, the hotter the gases get.

In **diesel** engines with extremely high compression, the mixture actually gets hot enough to self-ignite.

In **spark ignition** engines, the big bang of the power stroke is started by a small bolt of lightning across a spark plug gap inside the cylinder.

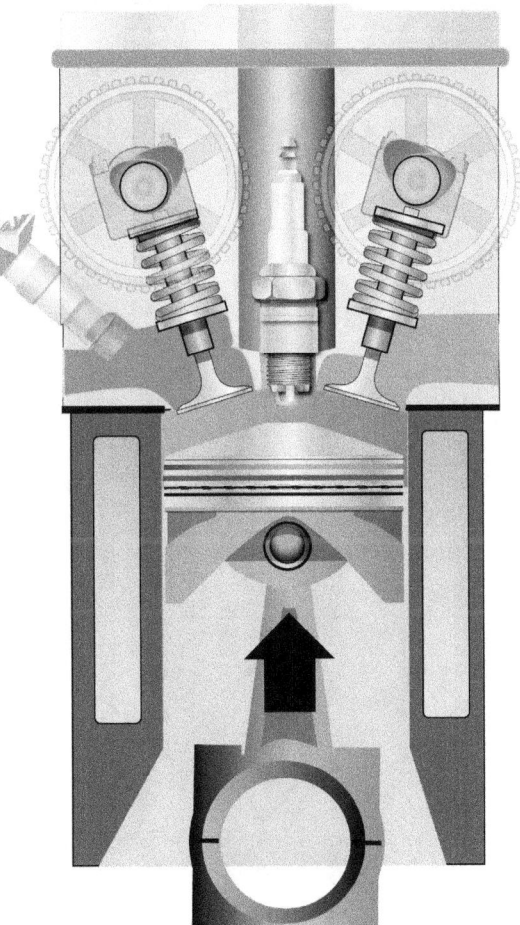

# Combustion

## Power Stroke

At the right instant, an arc of electrical energy jumps across the spark plug gap, igniting the compressed mixture. The force of the resulting explosion inside the cylinder depends on several factors, among them the ratio of oxygen to fuel, the degree of compression, the quality and octane of the fuel, and the timing of the spark event that starts the mixture burning.

When ignited, the gas charge expands rapidly as it burns, exerting great force against the only thing in the cylinder that can move—the head of the piston—driving it forcefully toward the crankshaft. It is similar to the force applied to a bicycle pedal by your foot: a linear force on the pedal rotates the chain sprocket.

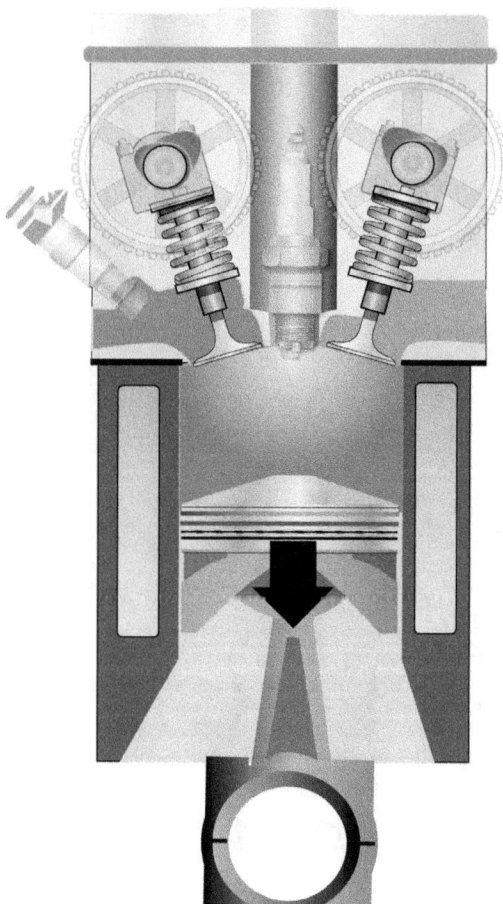

Several unwanted things can happen inside the cylinder that result in incomplete combustion, also known as **misfire**. Common misfire causes include a weak spark, incorrect spark timing, improper air/fuel ratio, poor fuel quality, and low compression.

See pages 132-133 for more about engine misfire.

The Auto Emissions Bible

## Exhaust Stroke

As the piston moves downward, the explosion burns out and its force weakens. Burned remains of the combustible charge are similar to the soot and ash left over after a wood fire goes out.

Any leftover pressure and waste gas material must be exhausted from the cylinder before a fresh charge of filtered air and fuel enters on the next intake stroke.

During the exhaust stroke, the exhaust valve opens and the next upward motion of the piston blows the waste gas out of the cylinder to the exhaust system.

This completes one entire four stroke combustion cycle. The process repeats itself as long as the elements needed for combustion continue to show up in the cylinders in the right proportions, at the right times.

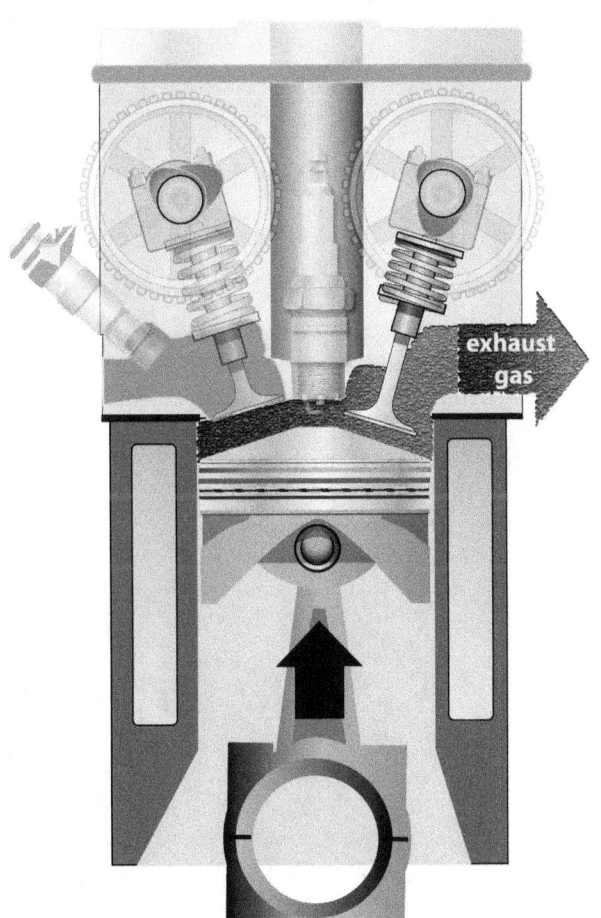

# Combustion (Oxidation Chemistry)

## A Chemical Reaction

So what exactly is this thing that we commonly refer to as *combustion*? It's a chemical reaction that combines fuel and oxygen to release energy by a process known as *oxidation*.

Along with water and all the gases mentioned so far, other combustion by-products include ammonia (NH3), sulfur dioxide (SO2), sulfuric acid (H2SO4), and aldehydes, ketones, carboxylic acids, soot and polycyclic hydrocarbons.

Fast oxidation creates a rapid expansion of burning fuel and oxygen that powers our vehicles. Burning gases push against pistons as a bicyclist pushes on pedals. The cyclist rotates a chain sprocket; pistons rotate a crankshaft. The cyclist and gas engine both oxidize fuel to do this work.

Let's think of an internal combustion engine as a device that creates conditions needed for the fast oxidation of hydrocarbons.

Here's the simplified general chemical equation for combustion:

$$O_2 + HC + heat \rightarrow H_2O + CO_2 + more\ heat$$

- Complete combustion is **clean** combustion. It produces energy, harmless water ($H_2O$), and Carbon Dioxide ($CO_2$), the same gas we humans exhale. New engine efficiency is impressive, but still less than prefect.

- **Dirty** combustion is incomplete burning that produces several unwanted gases, some of which are toxic and environmentally hazardous.

**targeted gases**

Unwanted gases include HC (unburned fuel), CO (partially burned fuel), and NOx (nitrogen and oxygen that combine at high combustion temperatures).

# Combustion (Stoichiometry)

## A Balanced Mixture

Chemical reactions like oxidation can take place successfully under good conditions—or fail when conditions are bad.

For every chemical reaction, there is an ideal mixture of ingredients that makes the reaction successful. This optimum mixture is referred to as *stoichiometry.*

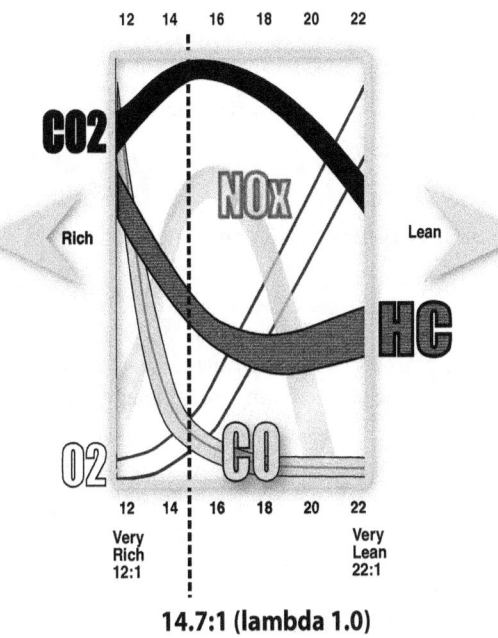

The stoichiometric conditions we need to look at here are for air and gasoline in an internal combustion engine. Stoichiometry for other fuels is different.

The stoichiometric air/fuel ratio for a compressed mixture of air and gasoline is about 14.7 to 1, by weight.

**14.7:1 (lambda 1.0)**

This ideal mixture is also called *lambda.* For more about this important concept, see pages 139-141 and 155.

---

## Why does it say "by weight"?

It takes 14.7 *pounds* of air to burn *one pound* of fuel, or 14.7 *tons* of air to burn *one ton* of fuel.

A gallon of air weighs far less than a gallon of fuel, so it takes about 9000 gallons of air to burn one gallon of gas.

---

# Combustion (by-products)

## Carbon Dioxide

Carbon dioxide ($CO_2$) is a non-toxic product of combustion. As a greenhouse gas, it is being targeted as a cause of climate change. As an indication of combustion and catalyst efficiency, however, more $CO_2$ is better.

exhaust gas

$CO_2$

**$CO_2$ exhaust levels increase with increased combustion and catalyst efficiency.**

- $CO_2$ is a product of complete combustion. High exhaust $CO_2$ indicates high combustion efficiency and a functioning catalytic converter.

- A good running engine with a properly functioning catalytic converter produces exhaust $CO_2$ concentrations of 15%-16%.

- Low $CO_2$ readings may indicate a mechanical fault. These commonly include problems like low compression, leaking valves, or improper camshaft or ignition timing.

- **Rule of thumb:** If $CO_2$ + CO percentages are lower than 14.5% (with AIR injection disabled), check the engine's mechanical condition first. If no engine mechanical issues are detected, move on to catalyst efficiency tests.

### Is that a tailpipe or a downspout?

If you burn a gallon of gasoline, you'll generate nearly a gallon of water. You may have seen the water dripping out of a cold tailpipe in the morning. When the water is hot enough, you can no longer see it, but water vapor is still a major component of exhaust gas emissions.

### OBD DTCs

**P0420-P0422**

**P0430-P0432**

Low Catalyst Efficiency DTCs may accompany low $CO_2$.

# Combustion (by-products)

## Carbon Monoxide

Carbon Monoxide (CO) is a toxic exhaust gas.

Carbon monoxide is produced by incomplete combustion. High exhaust CO indicates an air/fuel ratio imbalance with more fuel than air to burn it.

CO is a good indicator of a **rich mixture** (one containing too much fuel for the amount of oxygen available to burn it).

**CO concentrations of more than 1-2% measured upstream of the catalytic converter indicate a rich running condition.**

Possible causes of elevated CO include: excessive fuel pressure; leaking or dripping injectors; a faulty mass air flow sensor (MAF); a cold engine coolant temp sensor (ECT) reading; low $O_2$ sensor voltage, unintended fuel entering through a faulty purge control valve—in short, anything that dumps extra, unmetered fuel into the engine.

### CO: Odorless and Deadly

Breathing CO in concentrations as low as 300 ppm (0.03%) can be fatal in 30 minutes or less.

CO is odorless (exhaust odor is from hydro-carbons). CO is also a central nervous system depressant that makes us lethargic.

Every garage should have a functioning exhaust ventilation system. There should also be a CO detector with a digital readout and audible alarm running at all times.

DTCs commonly associated with high CO:

**P0172; P0175**
Fuel System Rich

**P0190; P0193**
Fuel Pressure Fault

**P0115, P0118 P0125- P0128**
ECT signal high or coolant temperature low; excessive time to enter closed loop

# Combustion (by-products)

## Oxygen

Oxygen is essential for combustion. The amount of oxygen left over after combustion can provide clues about combustion efficiency. For example, high concentrations of oxygen in the exhaust stream commonly indicate:

- **A lean mixture** (one that has too little fuel for the amount of air).

- **AIR system operation.** Temporarily disable any secondary air injection system, and ensure that the exhaust system is leak-free before sampling the exhaust gases.

- **Exhaust system leaks.** A leaking exhaust can draw air into the exhaust, diluting the test sample.

High $O_2$ readings may also result from: an air leak between the air flow sensor and throttle; low fuel pump pressure/volume; dirty or clogged injectors or a restricted fuel filter or fuel supply hose; a fuel injector power circuit voltage drop; or a contaminated air flow sensor. Cylinder misfire pumps a large amount of oxygen into the exhaust; the dead "hole" turns into an air pump.

DTCs commonly associated with high $O_2$:

### P0171; P0174
Fuel System Lean

### P0300 or P03xx
Engine random or cylinder-specific misfire

### P0410-P0419
Secondary AIR faults

# Combustion (by-products)

## NOx

**Oxides of Nitrogen** (NOx): NOx (pronounced "knocks") forms when the combustion chamber temperature exceeds 2500°F, the ignition point for nitrogen.

NOx has a slight odor and is reddish-brown in very high concentrations. It causes a stinging or burning sensation in the nose, eyes, and throat.

NOx should be considered as an indicator of high combustion chamber temperatures that commonly accompany high engine loads.

High NOx emissions may indicate: insufficient EGR flow; overly advanced ignition timing; high compression (from previous engine work or from internal carbon buildup due to a previous rich-running condition); cooling system faults; lean air/fuel mixtures; an inoperative knock sensor; low octane fuel; low fuel pressure or volume; etc.

See **Chapter 13: NOx.**

DTCs commonly associated with high NOx:

**P0127**
IAT High

**P0234-P0249**
Turbo/ Supercharger Faults

**P0324-P0334**
Knock Sensor Fault

**OBD DTCs**

**P0400-P0409**
EGR Faults

**P0480-485**
Cooling Fan Faults

NOx failures often occur right after vehicle repairs to correct a rich condition.

The leaner post-repair air/fuel ratio burns hotter. Carbon deposits left by the rich running condition jack up compression. Put the two together, and you have a perfect recipe for NOx.

# Combustion (by-products)

## Hydrocarbons

Hydrocarbons (HC): Remember that we're using HC to represent the many types of hydrocarbons found in raw gasoline.

HC in the exhaust stream is unburned fuel. It is present in the exhaust when it is not oxidized in the combustion process, in the exhaust system, or by catalyst action.

HC concentrations are usually measured and reported in parts-per-million (ppm). A good running engine with a properly functioning catalytic converter should emit fewer than 50 ppm of HC from the tailpipe. Many of the best running engines now exhaust 10 ppm HC, or less.

High HC concentrations are often associated with poor fuel economy.

Common causes for high exhaust HC:

- EVAP system vapor leaks to the intake manifold.
- A very rich or very lean mixture.
- Ignition misfire or poor fuel atomization. Check ignition timing and spark plug heat range.
- A leaking EGR valve.
- Improper cam or ignition timing.
- Worn plugs, wires, or other secondary ignition components.
- Uneven fuel injector delivery or dripping injectors.

An external fuel leak will cause high HC readings in the area of the leak. Make sure there's no raw fuel dripping near your emissions analyzer test probe.

## OBD DTCs

These DTCs are commonly associated with high HC:

### P0171; P0174
Fuel System Lean

### P0172; P0175
Fuel System Rich

### P0300; P03xx
Random or cylinder-specific misfire

### P0350-P0362
Ignition Coil/Circuit Faults

### P0400-P0409
Incorrect EGR position or flow

# Combustion (by-products)

## Review

This handy chart summarizes much of what we've learned so far. These are ideal values. Most vehicles won't have HC, CO, $O_2$ and NOx numbers this low, although some new, low emissions vehicles emit exhaust gases that are cleaner than the air they take in from the surrounding atmosphere!

| Gas | Engine Condition | Ideal Value |
|---|---|---|
| $CO_2$ | measure of combustion efficiency | >16% |
| CO | rich mixture, too much fuel, not enough air | 0 |
| $O_2$ | lean mixture, too much air, not enough fuel | 0 |
| $NO_x$ | high combustion temperatures, high engine load | 0 |
| HC | unburned fuel, leaking fuel or vapors | 0 |

## Combustion Review

- Fuel and air combine in the presence of heat to release energy.

- Our atmosphere is about 21% oxygen; 78% nitrogen, with trace elements making up the balance.

- Fuel is a soup of chemical compounds known as **hydrocarbons**.

- Combustion may be started by an electrical spark generated by a transformer, or by the heat of compression.

- Vehicle engines use a four stroke combustion cycle: intake, compression, power, and exhaust.

- The air/fuel ratio that provides the correct mass of air needed to totally combust a given mass of fuel is called **stoichiometry**. Engine inefficiencies prevent complete combustion, however.

- Gases monitored to determine engine efficiency include: $CO_2$, CO, $O_2$, NOx, and HC.

# THE MIL

# 2

## The MIL

The MIL (Malfunction Indicator Light)—also known as the Check Engine light—is the dashboard warning that lets the driver know when a Diagnostic Trouble Code (DTC) has been stored.

Here's how the light works in OBD II vehicles: Either it's ON with the engine running, indicating a problem in the vehicle that may increase harmful emissions—or it's OFF with the engine running, indicating that no serious emissions-related problems have been detected. (Some vehicles will also flash the MIL to indicate severe engine misfire.)

- The light should always come on at least briefly when the ignition is turned ON, but go out when the engine starts.
- If the light stays on with the engine running, or comes on while driving, it means that the onboard diagnostic software has stored one or more DTCs. The MIL alerts the motorist to the problem so it can be repaired.

Some motorists ignore the light; the emission test center won't. If the Check Engine Light (MIL) is ON with the engine running when your OBD II car is tested, it fails the test.

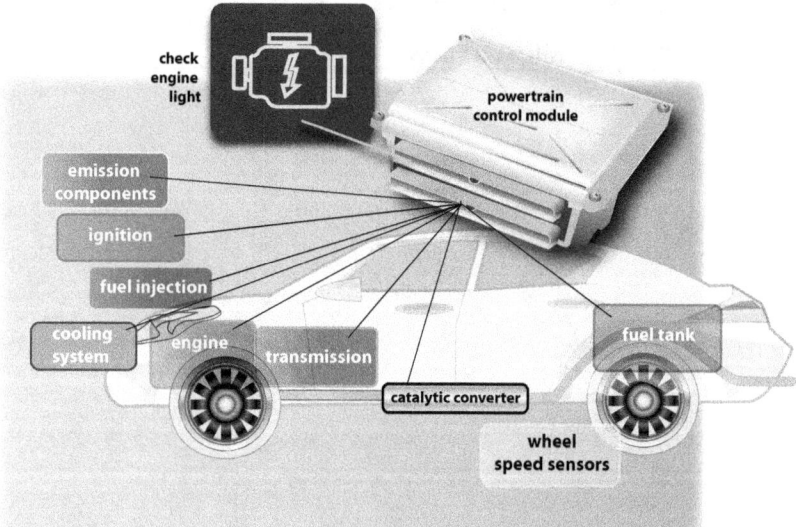

PCM diagnostic software evaluates sensor inputs from many vehicle systems and components. All tests are graded against an answer key. When test standards identify a failed system or component, the PCM stores a trouble code in computer memory, and turns on the MIL (Check Engine Light).

## OBD Sensors

### How Many Sensors Are There?

Your modern car has dozens of sensors; each one sends its report to the PCM while the vehicle is operated. That's a lot of data. Some tests performed by the computer run once per vehicle trip and then stop until the next trip. Others run *continuously*, as long as the car is driven.

### What If a Sensor Fails?

The PCM doesn't just monitor the car through the sensors, it monitors the sensors themselves. If a sensor signal gets whacky or disappears altogether, the PCM turns on the MIL (Check Engine Light).

### When Does the MIL Go ON?

If the PCM stores a DTC in memory identifying the sensor or system that has failed, it turns on the MIL. There are *thousands* of vehicle trouble codes; each one names a system or component and tells how it failed. Trouble codes stored in your car's computer provide clues about the nature of a MIL-illuminating vehicle problem.

Efforts to standardize naming conventions and other vehicle terminologies have been less that successful. Getting OEMs to agree on terms is like herding mice. Those of us on the receiving end dig through dictionary-length lists of terms. Many components have names and aliases: **MIL** and **Check Engine Light** are two names for the same thing; just one small example of OEM term-warfare.

# MIL

## Mil Illumination

The MIL comes on and stays on as long as the problem persists. Sometimes, the MIL will flash on and off if the problem is very serious and might cause vehicle damage.

There are two ways to turn off the light:

- **Use a scan tool command to turn off the MIL.** We do **not** recommend this procedure for the average motorist. Here's why: If you think you can make your car pass an emissions test simply by turning off the light with a scan tool just before the test, think again. The emission test computer will know what you did, and your car won't pass.

- **Fix the problem that turned on the MIL; then drive the car and let the PCM run its tests again.** If the PCM is satisfied the original problem is repaired (and it finds no other problem), it turns off the MIL (Check Engine Light). The car will then pass the emissions test.

An addition to OBD II strategies is something called the **Permanent Code**. Permanent DTCs can be erased **only** by the PCM. Even disconnecting the battery will not erase them. They are erased by the PCM only when onboard diagnostic tests confirm that the original fault that stored the DTC no longer exists.

## How Can I Retrieve Codes and Emissions Information?

Anybody can purchase and use a scan tool to view emissions data in OBD II vehicles. Many auto parts stores will also pull codes for free.

## Is It Safe to Drive the Vehicle With the MIL ON?

If the MIL comes on, find out why. Some problems may be simple and easy to correct. For example, a loose gas cap may turn on the MIL since the system detects a fuel system vapor leak. Other problems that turn on the MIL may be more serious and costly, if neglected.

## The MIL Matters

In its infancy, OBD II gained a reputation in some circles for being a little "cranky." Other words used to describe it during its awkward youth included: moody, balky, worthless, less-than-worthless, and several other shop terms too vibrant for this polite company.

As OEM engineers worked furiously to iron out its wrinkles, many motorists routinely ignored the MIL; others covered it with black tape if it came on. (It's lying anyway, right?) More ambitious technophiles pulled the bulb or cut its wire. Crude, Dude, but effective.

And so it was, that early on, with no penalty for doing so, MIL illumination was often viewed as a false alarm and ignored.

Mandatory emissions testing using the OBD II onboard data changed all that. Now, when your license plates are on the line, the MIL matters.

### OBD II Inconsistencies

Several OBD Modes are not used in many OBD II vehicles. Originally, OBD II scan tool connectivity was a hit and miss affair. Some vehicles actually got "waivers," the equivalent of a Mulligan on the golf course, when their software failed to work properly. One manufacturer chose to disable the misfire monitor in some models, preferring to pay a fine and compensate vehicle owners instead. More of these issues were solved with each passing model year, but there are still plenty of OBD II vehicles on our streets that do not work and play well with others.

# MIL

## OBD II Scan Tool Emissions Tests

Gradually, OBD II software has improved. In fact, its reliability has been deemed so good in recent model years, that many states use a vehicle computer scan as a replacement for tailpipe testing on all OBD II vehicles.

When evaluating this change, most states found that the advantages of **Scan Tool Emission Tests** were too great to ignore:

• The test equipment for a scan test is WAY cheaper than the cost of Titanic-sized, 10-ton emissions analyzers needed for tailpipe testing.

• Scan Tool tests are faster, cleaner, and less dangerous than running a couple tons of vehicle on a dynamometer—inside a building.

• It's easier to sell the motoring public a fast, quiet, hi-tech scan. People trust computers.

• Tailpipe test accuracy depends on several variables that called their accuracy into question. Vehicle operating temperature and the skill of the employee behind the wheel during the test could both have a big impact on the final results.

• Test center employees use a bar code scanner to gather critical vehicle data, including the Vehicle Identification Number (VIN). Some cars provide their VIN as part of their stored emissions data. Fast, error free data collection moves cars through test lanes quickly, and reduces manual data entry errors.

• All vehicle data can be sucked out of the vehicle computer and piped to a mainframe miles away over phone lines, untouched by human hands.

• It's cheaper. Okay, so we repeated that one.

## Scan Tool Tests and the MIL

If you live in an area where emissions tests are done with a Scan Tool, it's pretty hard to ignore an illuminated MIL. **MIL status passes or fails a vehicle in the first stage of the Scan Tool Emissions Test.**

This is no small deal, ladies and gents: in some places, if your car fails, you don't get new license plates until the vehicle is fixed and passes a retest; and **retest standards may be tougher**.

### First, the MIL Test

When you pull into the vehicle emission test center, the first thing they check is the MIL.

- The MIL must come on when the ignition is first switched to the ON position. This is called a *bulb check*, and it's used to confirm that the bulb is still in the dash—not in your tool box or pocket—and that it is being commanded by the vehicle computer.

- The MIL must go out and *stay out* when the engine starts.

- If the MIL doesn't illuminate key-on, and then go out when the engine starts, the vehicle fails the test—period. There is no room for interpretation.

KOEO stands for Key-On-Engine Off

KOER stands for Key-On-Engine-Running

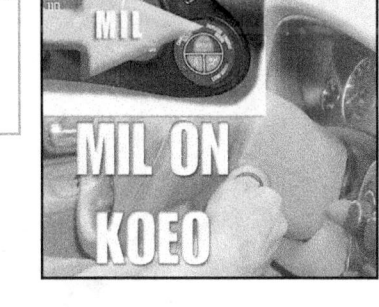

You perform these tests every time you start your car. Just make sure the MIL comes on with the key in the on position, and that it goes off when the engine starts.

## MIL Tampering

We Americans are a resourceful lot, and many back yard inventors have tried to game the OBD II system over the years. Some folks always feel that cheating to pass the test is easier than fixing the car.

- **A few creative characters have tried to fool the Scan Tool test by wiring the MIL to the oil pressure switch.** Think about it: the oil light comes on key-on-engine-off, and then goes out when the engine starts. (That's the creative spirit that made America great!)

- **Others have tried to use a good car to get a broken car through the test.** Several test station strategies are now used to prevent this, and newer models can be identified by scanning their Vehicle Identification Number (VIN). (It's like a pro athlete juicer who sends in a blood sample donated by his steroid-free brother.) Matching the vehicle registration and VIN provide fingerprint-positive vehicle identification in later model vehicles to prevent this activity.

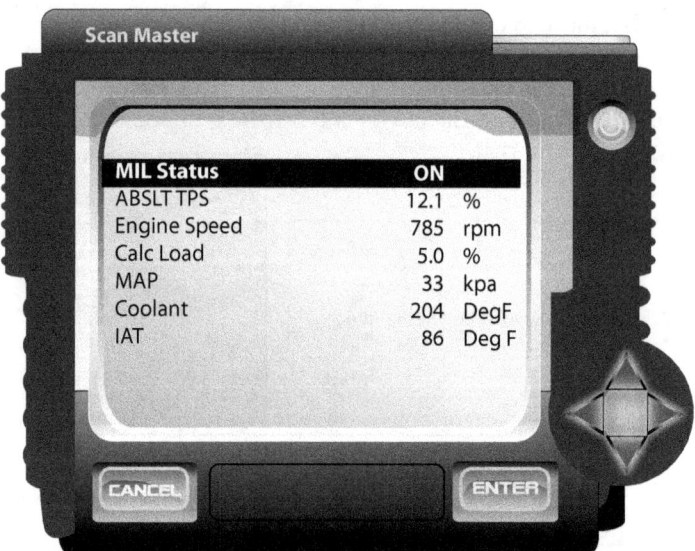

The visual MIL check is followed by a scan of the vehicle computer. The command sent by the computer to the MIL **must** match its actual state (ON or OFF). If the MIL command says ON (as it is in our image), but the MIL is not illuminated with the engine running, the motorist has some explaining to do. That's why MIL status shows up in datastream.

The Auto Emissions Bible

## MIL Fun Facts

The vehicle computer, referred to as the Powertrain Control Module (PCM), controls the MIL.

- **The PCM turns ON the MIL.** The MIL illuminates when a DTC is stored in the PCM. Additional DTCs do not change MIL illumination after the first DTC is recorded. The MIL stays on until it is turned off by the PCM or by a command from the scan tool to erase DTCs. New, **Permanent DTCs** can be turned off only by the PCM.

- **The PCM can turn OFF the MIL.** The OBD II test that turned on the MIL initially must run and pass three times for the PCM to turn off the MIL.

- **Remove battery power from the PCM long enough, and all vehicle data stored in the PCM is erased, including DTCs.** This turns off the MIL and resets all non-continuous monitors to *not ready* (aka *incomplete*). (**Permanent DTCs** cannot be erased by removing PCM power; they are stored in non-volatile memory. Only the PCM can erase them)

- **You can turn off the MIL with a scan tool by erasing DTCs.** Unfortunately, this also erases all test results and emissions-related data stored in the PCM. The vehicle will not pass a scan tool emissions test in this condition. It must be driven to complete those tests again before it can be tested. (Again, this does not apply to Permanent DTCs.)

## MIL Review

- The PCM turns on the MIL when a DTC is stored.

- The MIL should come on at least briefly when the ignition switch is turned to the ON position, and then go out when the engine is started.

- The MIL is also referred to as the Check Engine Light.

- The MIL may be turned off (at least temporarily) by a scan tool command to erase emissions data, or extinguished by the PCM, if onboard tests indicate that the DTC-storing fault no longer exists.

- MIL illumination was often ignored prior to scan tool emissions tests linked to license renewal.

- The MIL test is now a central part of many vehicle emissions tests.

- Permanent DTCs can be erased only by the PCM.

# DTCS
## DIAGNOSTIC TROUBLE CODES

# 3

# DTCs

## Sign Posts

When the MIL comes on, OBD II doesn't just leave you guessing; it puts up sign posts that point you in the right diagnostic direction.

Those sign posts are called **Diagnostic Trouble Codes (DTCs)** and there are thousands of them. How many exactly? I have no idea, and there is no sense counting them; more have been added since I wrote this, and there are more on the way.

New components and new technologies demand new labels.

Expect the DTC list to grow like weeds in a pea patch. Hybrid-electric vehicles, gas direct injection, and hi-tech diesel exhaust systems have each added entire series of numbers to the master DTC list. Tighter emissions standards and new stricter fuel efficiency standards are generating new technologies, daily.

## The First Diagnostic Step

The Malfunction Indicator Light (MIL) in your car is illuminated. (This is the warning lamp also referred to as the Check Engine Light).

Our first diagnostic step is to check the vehicle computer for codes. The easiest way to look for codes is with a scan tool, although some cars display codes on the dashboard odometer or driver information center. Sometimes, we'll find one DTC; sometimes we'll find several.

When we "pull" codes from the vehicle computer (PCM), we ask the scan tool to send a text message saying, "Yo, computer: send me a list of DTCs, willya?" The computer responds by firing off the data, which is then displayed on the scan tool.

Diagnostic Trouble Codes are the first computer troubleshooting tool we learn to use, and the one we are generally most comfortable with. Knowing how to use DTCs properly is another matter entirely.

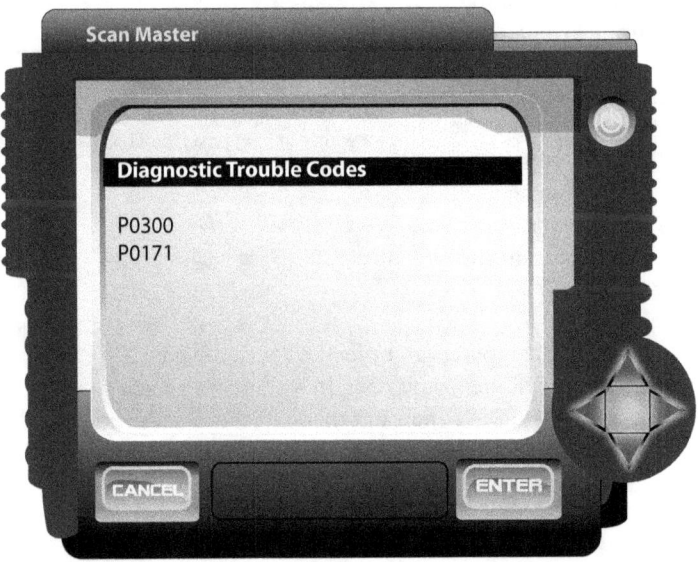

DTCs break the ice. They send us in the right general direction. They are not a diagnosis, only a sign post.

# DTCs

## DTC Numbers

Each OBD II DTC "number" is a five-character label, made up of both letters of the alphabet and numbers. As a result, DTCs are said to be *alphanumeric*.

### A Sample DTC

Here's a common DTC: **P0300: random engine misfire**

**P0300** is so common that professional repair technicians know it by heart. It indicates that there is emissions-related misfire in one or more of the engine cylinders. The exact location of the misfiring cylinder or cylinders is not given by the DTC number P0300.

Clearly, the DTC is pointing us in the right general direction, but it's up to us to locate the exact location and cause of the failure.

---

**Google That DTC**

Google "**P0300**" and you'll get a long list of links to various web sites, some that list generic DTCs, and others that cater to those looking for just-in-time repair information.

The more common the DTC, the more hits you'll get.

In fact, you can Google almost any DTC and get some feedback that may even include a detailed explanation of the affected circuits, and trouble-shooting advice.

Some links take you to forums where participants discuss personal repair experiences. Like all internet information, you'll need to exercise caution when selecting advice.

Some bloggers are reliable; others, less so. A few will recommend procedures that are ineffective, stupid, or downright dangerous. Play it safe: cross reference all internet information, and compare procedures in blog posts to published repair standards from reputable sources.

---

**Note:** Some vehicle computers go beyond a general misfire DTC and identify a misfiring cylinder by number. In fact, newer vehicles are required to do so. For example, P0302 indicates that a misfire is occurring in cylinder number 2.

---

## DTCs are sign posts

---

## Code Number Classes

The graphic on this page shows how DTC numbers are assigned.

Relax, you don't need to memorize this information. In actual practice, you'll pull vehicle DTCs from the vehicle, write them down, and look them up in a DTC reference book, repair database—or Google!

The numbering system below applies specifically to **Powertrain** DTCs (those beginning with the letter **P**).

For **Body (B)**, **Chassis (C)**, and **Network Communications (U)** codes, a second digit of 2 is reserved for OEM DTCs.

The **P3xxx code numbers are split**; those from P3000-P3399 are OEM reserved; those from P3400 to P3999 are SAE reserved.

The second digit indicates whether the DTC is defined by the Society of Engineers (SAE) or by an OEM.

A second P-code digit of 0, 2, or 3 (3400-3999) indicates an SAE- defined DTC

A second P-code digit of 1 or 3 (3000-3399) indicates an OEM-defined DTC.

The **first digit** is a letter of the alphabet that indicates the main system where the fault occurred.

**P = Powertrain**

**B = Body**

**C = Chassis**

**U = Network Communications**

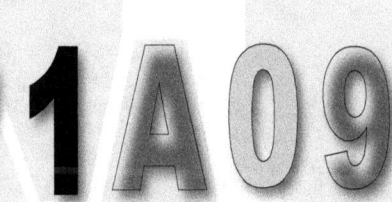

The **third** digit indicates the affected subsystem.

• P0Axx - **Hybrid**

• P00xx - P01xx - P02xx - **Fuel and Air Metering**

• P03xx - **Ignition and Misfire**

• P04xx - P05xx - **Vehicle Speed, Idle Control, and Auxiliary Inputs, EGR, EVAP, AIR**

• P06xx - **Computer and Auxiliary Inputs**

• P07xx, P08xx, P09xx - **Transmission**

The **fourth and fifth digits** are a fault number assigned to the component and fault recorded:

Common examples of fault types include:

• circuit fault

• circuit low

• circuit high

• circuit range/performance

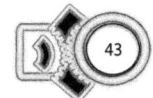

# DTCs

## Where Do DTC Numbers Come From?

Two sources:

- **The Society of Automotive Engineers (SAE).** SAE-defined DTCs are sometimes called *generic* codes, since they apply to systems and components used in a broad range of vehicles.

- **Vehicle makers.** OEM-defined DTCs generally apply to components and systems that are unique to specific makes or models, although some insist on applying their own number to a generic component.

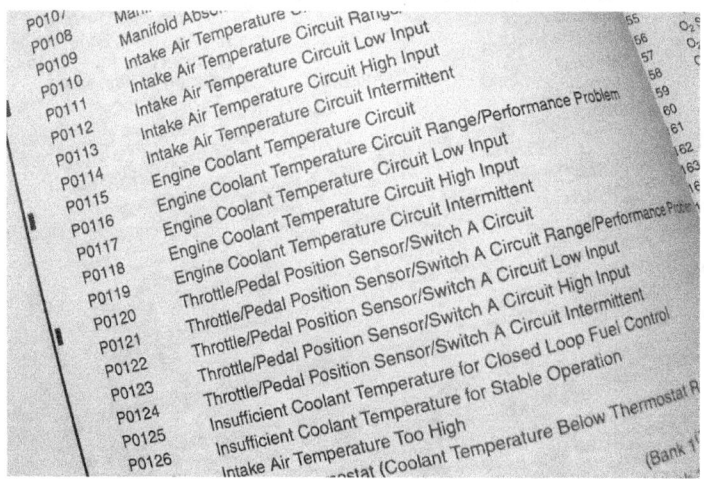

This is a page from SAE paper J2012. It shows DTC numbers listed by SAE, followed by their descriptors.

All SAE papers can be purchased from the SAE book store at *www.sae.org*.

DTCs are sign posts.

Descriptors are labels.

## What's In a DTC?

If you look at a master list of DTCs, you'll notice that many common components have several DTC numbers assigned to them.

Let's look at an example showing how a single component can store one of several DTCs, depending on the nature of the failure.

### DTCs for a Common Sensor

The **Engine Coolant Temperature Sensor (ECT)** is found in all modern vehicles. ECT resistance decreases as engine coolant temperature increases, changing ECT circuit voltage. The ECT signal is essential for many vehicle computer functions. The PCM looks at ECT when it adjusts fuel delivery and ignition timing. ECT signals also control the electrical cooling fans and detect engine overheating.

DTCs listed below show us that the ECT and its circuit can fail in several ways. Each DTC number is labeled by failure *type*. Since the ECT is a component common to vehicles from various manufacturers, these DTC numbers are "generic." Their numbers and descriptors are assigned by the Society of Automotive Engineers (SAE). Generic DTC numbers may be used by **any** OEM to identify common ECT circuit faults.

**P0115** - Engine Coolant Temperature Circuit Malfunction
**P0116** - Engine Coolant Temperature Circuit Range/Performance
**P0117** - Engine Coolant Temperature Circuit - Low Voltage
**P0118** - Engine Coolant Temperature Circuit - High Voltage
**P0119** - Engine Coolant Temperature Circuit - Intermittent

### Descriptors

See how each DTC number is identified by a short label? We'll call these labels *descriptors*. "Engine Coolant Temperature Circuit - Low Voltage" is a *descriptor*. It names the component and fault condition, but offers no details.

Descriptors are NOT definitions.

# DTCs

## What Descriptors Tell Us

DTC numbers and descriptors are sign posts, not exact addresses. It's like the difference between, "they went thataway" and, "they went to 240 Elm Street."

Let's run down the list of generic codes for the ECT.

- **P0115** tells us only that an undefined fault has been detected in the ECT circuit.

- **P0116** gets a little more specific, but not much. Apparently, the ECT signal is out of range, or its performance is suspicious.

- **P0117** shows one extreme, a sensor whose voltage is too **low**; below the test limit. This usually indicates that there's a short circuit to ground between the PCM and sensor.

- **P0118** is also more specific than P0115: sensor signal voltage is **high**. Its voltage or frequency is greater than the test limit. This usually indicates that there's an open in the sensor or its circuit. If P0117 is "heads," P0118 is "tails."

- **P0119** suggests an intermittent condition in the ECT circuit. Again, this is pretty vague.

### Troubleshooting Tip

Many inexperienced mechanics automatically replace any component mentioned by the DTC descriptor. Be careful with this approach: loose or corroded connections or damaged wiring are very often the underlying cause of DTCs.

For example, while a **P0117** might indicate a bad ECT sensor, the descriptor clearly indicates a problem in the ECT "circuit." The ECT sensor is certainly part of the ECT circuit, but it is not the *entire* circuit.

When diagnosing a circuit, include the sensor connector, sensor circuit wiring, the PCM, the sensor ground, etc., as well as the sensor itself. This applies to all sensors.

Savvy diagnosticians use the DTC to point them in the right direction, and then use sound diagnostic techniques to home in on the exact cause of the problem. (Remember the tests on page 10?)

## SAE vs. OEM DTCs

Let's idle by the curb for a minute here and talk more about the difference between SAE-defined DTCs and OEM-defined DTCs.

If you look up a **P0117**, it will have the same *descriptor*, whether it's used for a Hyundai or a Hummer. The second character—a zero—tells us it's generic: it applies to a common component found on *many* cars.

OEM-defined DTCs are very different. If Zombie Motors uses a reverse-widget-gravitator not found in any other brand, they will choose an OEM (manufacturer-specific) DTC number. The component is proprietary.

### Look It Up to Be Sure

Unlike generic DTCs, OEM-assigned DTC numbers have very different descriptors. We grabbed a random OEM-DTC number and looked up its descriptors for different vehicles. Clearly, a P1107, manufacturer-specific DTC means different things to different OEMs.

**P1107 - Subaru** - AIR System Diagnosis Solenoid Circuit Failed
**P1107 - Honda/Acura** - Barometric Pressure Sensor Circuit Low Voltage
**P1107 - General Motors** - MAP Sensor Intermittent Low Voltage
**P1107 - Isuzu** - MAP Sensor Intermittent Low Voltage
**P1107 - Audi** - $O_2$ Sensor Heating Circuit, Bank 2-Sensor 1, Short to B+
**P1107 - Ford** - Dual Alternator Lower Circuit Malfunction (Powerstroke Diesel)
**P1107 - Ford** - Manifold Absolute Pressure (MAP) Sensor Circuit Intermittent Low Voltage

Here we see that a single DTC number—**P1107**—has been used by multiple OEMs to designate very different faults.

---

## It pays to look up DTC definitions.

---

# DTCs

**Heads Up!**
This next part is very important. To make and confirm repairs, you need to know exact test standards and limits.

## DTC Descriptor vs. DTC Definition

How does a vehicle manufacturer choose one DTC number out of thousands to describe a fault? We can't be sure, can we? That stuff all happens behind the curtain back in Oz.

An OEM could conceivably slap a P0115 (Engine Coolant Temperature Circuit Malfunction) on *all* ECT circuit problems. (Heck, "malfunction" could mean anything.) This is frowned on.

Ideally, for low circuit voltage, they'll go with a P0117. Do we know more with the P0117 than we did with the P0115? Some. But I am hungry for more information that will help me fix the car.

- **I want the exact test limits.** I want steak and potatoes test values that I can measure! Does a sensor fail at 3.5 volts or at 5.0 volts? Give me test standards with hair on their chests, expressed in volts or frequency; units of pressure or degrees of temperature—things I can measure with normal shop equipment.

- **I want to know the exact vehicle operating conditions required to run an onboard test.** Is a *frumple sensor* tested as soon as the ignition is switched on? Two minutes after the engine is started? When driving at 35 mph, or faster? I want to *know* these things so I can test under similar operating conditions.

- **I want to know if there are special operating conditions needed to run an onboard test.** Will the test be suspended (delayed) during some driving conditions?

I want more information than the code number and ***descriptor*** give me. I want a ***definition***, chock full of test data. Vehicle repair is a science experiment that demands accuracy; it ain't horse shoes! If somebody sends me driving directions that say: "I live East of Chicago," I won't leave the house until I call back and get an exact street address. **Descriptors** are sign posts; **Definitions** should provide us with GPS coordinates!

## DTC Definitions

Here's what a well-written DTC definition does:

- **It lists the conditions needed to run a test.** Is it enough to have the key on for a test to run? Does the engine have to be running? Do we need to be traveling down the road at a certain speed?

- **It spells out the test limits.** Test limits are the answer key for each onboard test. All pass-fail test limits should be expressed either as a calculated value (e.g. degrees of temperature), or as voltage.

 Experience with real world OEM repair information tells us that some DTC definitions are better written than others. Good ones are rich with test data; others are thin on the details, and leave us guessing.

### Where Do DTC Definitions Come From?

Each time an OEM assigns a DTC number, it defines the DTC.

Once a DTC is assigned to a vehicle, it is no longer generic.

Even a generic, SAE-defined DTC number and descriptor have a vehicle-specific **definition** once they are assigned. The **descriptor** is generic; the **definition** is specific to the vehicle application.

That's why we keep insisting that there is a difference between *descriptor* and *definition*.

> **SAE paper J2012 - Diagnostic Trouble Code Definitions**, establishes the numbering conventions for OBD II DTCs.
>
> Oops, there's that word "definitions" again. Not the best choice. While J2012 certainly contains a list of generic DTCs with *descriptors*, it certainly has no code *definitions.*
>
> That's because DTCs get their definitions only when they are assigned to a vehicle by a manufacturer, and not before.

# DTCs

## DTC Definition Examples

A well-written DTC definition is the place to look for monitor test conditions and pass-fail standards.

The following examples are a partial list from GM for an SAE-defined P0117. All are from the same model year, but notice that the DTC definition changes as P0117 is applied to different models.

These definitions all apply to the same DTC number (**P0117**) and share the same descriptor (**ECT Sensor Circuit Low Voltage**); but the details spelled out in the definitions tell us they are **not** all the same.

We repeat: These are all from one OEM, applying different standards to a variety of models.

 Three test limits are given in degrees of temperature; while one lists a test voltage standard. Two require a minimum engine run time; two do not. One size does not fit all.

- **After the engine has run for at least 128 seconds**, the PCM detects an ECT input of **more than 289°F**.

- **Engine run time over 15 seconds**, the PCM detects an ECT input of **over 284°F**; the condition **lasts for at least 15 seconds**.

- Code conditions: **This code may be set only if there are no other codes already stored in memory.** The PCM will store a P0117 if it detects an ECT sensor input **greater than or equal to 304° F**, either condition **met for 2 seconds**.

- **Engine running** the PCM detects an ECT voltage input of **less than 0.78V for 5 seconds**.

If we want to know which measurement standards to use during testing, we need to look up the DTC definition or locate the test standards in a repair database.

The Auto Emissions Bible

## One-Trip DTCs

A **one-trip** DTC is stored by a One-Trip Fault. Expect this to be used with clear-cut faults like shorted and open circuits in critical engine sensors. The PCM sees the faults, stores a DTC, and turns on the MIL; in one trip.

## Two-Trip DTCs

**Two-trip** faults must be detected on a second trip to illuminate the MIL. Two-trip fault detection is used to prevent false MIL illumination. Some monitors will run more than twice when testing complicated systems. Catalyst and EVAP leak tests are examples.

## Pending DTCs

A pending DTC is stored in memory and may be displayed on the scan tool to indicate a failure during the first trip of a two-trip fault. It is a warning sign, and an indication of a failed test that will mature into a full DTC, should it be detected again on the next trip.

When properly implemented by an OEM, pending codes show us a developing problem or alert us to the *persistence* of a problem, even after a repair. In either case, **Pending DTCs** are stored in a single trip. They tell us that the vehicle has an issue that may mature into a MIL-illuminating DTC. Think of it as a DTC *early warning system*.

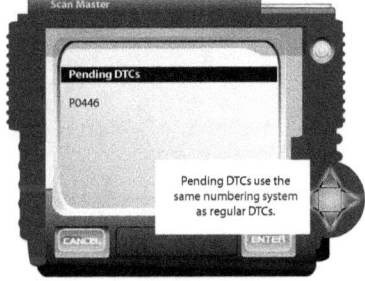

Some OEM diagnostic procedures even suggest erasing DTCs and then running the monitor that stored the pending code again by completing a trip. If the original fault persists, a pending code may be stored, telling us to go back to the drawing board.

Vehicle makers may also use "Check Mode," a special diagnostic function that temporarily *converts all DTCs to one-trip codes*. This feature is activated by a scan tool command or manually, with a jumper wire inserted into a diagnostic connector. Check vehicle service information for availability and correct procedures.

## Permanent DTCs

To prevent cheating, a new type of DTC has been added, called a **Permanent DTC**. Unlike previous DTCs that can be erased with a scan tool (or by removing battery voltage from the PCM), the Permanent DTC is stored in an area of PCM memory that is not erased when power is removed. Expect this change to be phased in between 2010 and 2012. Permanent DTCs will be identified by a unique label to identify them.

# DTCs

## Freeze Frame

Skilled auto repair professionals have long used their scan tools to make data recordings. Recordings can be triggered manually, by a user-defined event, or by a DTC. Multiple frame data recordings are called *movies*. A single frame from that movie is referred to as a **snapshot**.

- OBD II stores its own single frame data snapshot automatically, when triggered by a DTC. **Freeze Frame** is vehicle data recorded at the time the DTC is stored. It is a special diagnostic feature built into OBD II vehicles.

- The OBD II standard requires that at least *one* Freeze Frame be stored, usually with the first DTC, although more than one Freeze Frame is available in some models. If a more serious DTC occurs after the original DTC, Freeze Frame may be overwritten by data from the newer code.

- All Freeze Frame data are stored with the DTC until both are erased from the vehicle computer memory.

- The amount of data you get from a Freeze Frame varies. Whether or not it is useful, depends on the nature of the problem and the age of the vehicle. Early Freeze Frame is limited; recent vehicles are more generous with diagnostic data, listing many more parameters.

| Code | Description | | | |
|------|-------------|--|--|--|
| | DTC Codes: | | | |
| P0481 | Fan 2 Control Circuit Malfunction | | | |
| | No Pending Codes Present | | | |

| Supported PIDs | | Abbrev | Data | Units |
|----------------|--|--------|------|-------|
| P0481 DTC caused Freeze Frame Storage #0: | | | | |
| Calculated Load | | LOAD_PCT | 4.7059 | % |
| Engine Coolant Temperature | | ECT | 85.0000 | Deg C |
| Short Term Fuel Trim Bank 1 | | SHRTFT1 | 0.0000 | % |
| Long Term Fuel Trim Bank 1 | | LONGFT1 | 0.0000 | % |
| Intake Manifold Absolute Pressure | | MAP | 31.0000 | kPa |
| Engine RPM | | RPM | 1386.5000 | RPM |
| Vehicle Speed Sensor | | VSS | 21.0000 | km/h |
| Air Flow Rate from Mass Air Flow Sensor | | MAF_g/s | 7.5200 | g/s |
| Air Flow Rate from Mass Air Flow Sensor | | MAF_lb/m | 0.9926 | lb/m |
| Absolute Throttle Position | | TP | 3.1373 | % |

Be careful with Freeze Frame. Sometimes it's useful. Sometimes it isn't. Early Freeze Frame concentrates heavily on engine performance PIDs; data categories that may not provide useful info on non-engine codes. This Freeze Frame for a cooling fan fault in a GM vehicle does not offer useful diagnostic clues for a P0481. (See page 241 for the fix.)

## Putting DTCs to Work

A DTC is a tool. Like any other tool, its usefulness depends on the skill of the user. Keep the following in mind whenever you use DTCs:

- **The PCM occasionally points the finger of guilt at an innocent victim!** A transmission DTC may be stored if the Transmission Control Module (TCM) fails to see a critical sensor signal input like the vehicle speed or throttle position sensor.

- **Some vehicle problems will not store a DTC, even if the problem is emissions related.** For example, a component that is binding or sticking mechanically—but passes the electrical test—may not store a code.

- **A vehicle with a DTC stored in memory may have no symptoms**; this is more common than you think.

  **Example:** A gutted catalytic converter will store a DTC: the catalyst isn't cleaning the exhaust, but the car will run perfectly.

  **Example:** An evaporative emission system vapor leak (maybe from a loose gas cap?) may illuminate the MIL, but have no effect on vehicle performance.

  Sophisticated PCM fail-safe strategies keep many vehicles moving down the road, even when something serious is broken. The PCM takes over and "plugs in" a sensor value to keep the engine running. For example, the PCM may substitute the air intake sensor signal if it cannot see the signal from the engine coolant sensor. (For more on data substitution, see pages 88-91.)

- **Some symptoms won't store DTCs.** The flip side of the previous coin is that OBD II will occasionally run some active tests that cause brief "symptoms." These may seem like a problem, but are not. For example, an onboard self-test strategy may produce a brief rolling idle while fuel vapors are purged from the charcoal canister into a running engine. This is normal in some vehicles. Don't rush to fix something that is not broken!

- **DTCs may not tell us the entire story.** An open circuit in a shared electrical circuit may affect several sensors, even though only one fault is identified by the DTC. Problems in shared voltage supply and ground circuits can affect many components at the same time.

# DTCs

## DTC Cautions and Pitfalls

- **Some DTCs identify a *system* problem — not a faulty component.** You'll have a deuce of a time diagnosing a low purge flow DTC without some knowledge of how the purge system works. To correct a low purge flow condition, you first need to know what purge flow means. Then you need to know which components control purge flow, when they operate, how they are controlled, and where they are located.

- **A vehicle with multiple faults may not store multiple codes.** OBD II is designed to reduce the old problem of cascading faults. This was a lot bigger problem in previous onboard test systems when a single fault started an avalanche of codes. To prevent cascading DTCs, one or more OBD II monitors may stop running altogether when a DTC is stored. With some monitors asleep, additional faults are not detected. To get the monitors running again, correct the original fault, and erase DTCs.

- **A vehicle with multiple DTCs may have a problem unrelated to any of the codes found in memory.** For example, a voltage drop in a common sensor ground connection can disrupt multiple circuits. This drives the PCM bananas, and it misdiagnoses the problem.

---

**One Fault = Multiple Codes**
A single fault in an electrical circuit can drive several components batty at the same time.

**Example:** 2003 Chevrolet Suburban 1500 5.3L
**Symptom:** When driving over bumps, the Anti-lock Brake System (ABS) light and Red brake light come on. The MIL may flicker; power door locks lock and unlock; the chime sounds.

The following DTCs are stored:
- **C0237** stored in the ABS system.
- **P0706** stored in the Powertrain Control Module (PCM).
- **B1000** stored in the Transfer Case Control Module (TCCM).
- **U1000** stored in the Instrument Panel Cluster (IPC).

**Confirmed Fix:** Tighten or repair loose or damaged ground wires at the rear of each cylinder head (grounds **G103** and *G104*).

(Our thanks to **Identifix** - www.*Identifix.com* for this case study from their hotline archive.)

---

## DTC Cautions and Pitfalls (cont)

- **DTCs in the hands of the novice can result in serious injury (or worse).** This is particularly true of hybrid vehicles where *lethal* voltages may be encountered. Hybrid vehicle faults involving high voltage DTCs should always be referred to a properly trained professional who has access to recommended safety and test equipment, and vehicle-specific documentation.

This **Hybrid Electric Vehicle battery packs 275 attention-getting, heart-stopping volts**—enough to seriously injure or kill an unsuspecting human who sticks his finger in the wrong place. We **strongly** urge untrained individuals to leave diagnosis and repair of high voltage hybrid vehicle systems to properly trained and equipped professional technicians.

- **DTC symptoms may be difficult to predict in complex systems.** This is especially true where network communication codes are involved. Many modern cars have several computers connected over a network. Network problems store DTCs that start with the letter "U."

**U-code faults can be hard to find.** The skill level required to isolate a U-code fault depends on:

- the number of networks installed in the vehicle.
- how the networks are connected.
- whether the fault is still present.
- whether the failure occurs with the engine off or with the engine running.

Like hybrid vehicle faults, network troubleshooting commonly requires special test equipment, exact vehicle repair data, a lot of patience—or all of the above.

# DTCs

## Keep Things Simple

Play the odds. Seek simple fixes for common problems instead of wasting hours looking for unlikely solutions. The KISS rule: Keep It Simple, Stanley, is always a high probability choice. Most of the time, a DTC will point you in the general direction of your problem.

Let's say you pull a P0118 (engine coolant temp input high). Freeze Frame data for the DTC records a -40° ECT temperature. Odds are, you'll find a bad ECT or a wiring fault. (If you hear hoofbeats, think horses, not zebras!)

### Exceptions to the Rules

K.I.S.S. is a high probability play, not a guarantee. Stuff happens. Most of the time, OBD II makes the right call; sometimes it makes a mistake, and blames the wrong suspect.

Here are a few real world problems seen often enough in Sam's shop to be labeled as repeat offenders:

- **A Toyota with a P0125** (Insufficient Coolant Temperature for Closed Loop Fuel Control). This one sure sounds like a vehicle that isn't reaching normal operating temperature, doesn't it? Probably a bad thermostat, right? Not necessarily. Every time this fault has shown up at the shop, the real cause has been a faulty $O_2$ sensor (or a faulty air/fuel sensor, in Toyota vehicles that have them).

- **A Toyota with two DTCs. P0441** (Evaporative Emission System Incorrect Purge Flow) and **P0446** (Evaporative Emission System Canister Vent Solenoid Fault). Odds are high that the problem lies in the EVAP Bypass VSV (Vacuum Switching Valve), which the PCM uses to switch the Fuel Tank Pressure Sensor from reading tank pressure to reading canister pressure.

- **A Toyota with a P0401** (Insufficient EGR Flow). Test for EGR control vacuum during a road test. Then listen for a noticeable rough running condition while manually opening the EGR at 1800 rpm. If both these tests pass, replace the EGR VSV or solenoid. The real problem isn't insufficient EGR flow, it's that the flow *doesn't start soon enough*.

The Auto Emissions Bible

## Exceptions to the Rules

- **A GM 3.1L minivan with a P0171** (Fuel System Lean, Bank 1) may really be running lean, but based on lies told to us before by these vehicles, our money rides on a sluggish oxygen sensor.

> To check the $O_2$ sensor response, add propane at the air cleaner housing inlet, and expect to see $O_2$ voltage rise. If the voltage doesn't increase enthusiastically, stick a fork in the $O_2$ sensor and replace it: it's cooked. Of course, if you have access to an exhaust gas analyzer, check lambda. See pages 139-141 and 155 for more on lambda.

- **A Volvo with a P0118** (ECT Sensor Circuit High Voltage). This is a generic DTC indicating a high ECT signal input. This is one case where Freeze Frame can be a big help. If the ECT generic data PID is anything except -40°, the cooling system thermostat is probably bad! It's hanging open, keeping the engine too cool for too long; the ECT input is higher than expected, hence the code choice.

- **Fords with a P0401** (Insufficient EGR System Flow Detected). Flow is actually okay. The cure? Nine times out of ten, a new DPFE sensor is the correct fix. The old one isn't detecting flow.

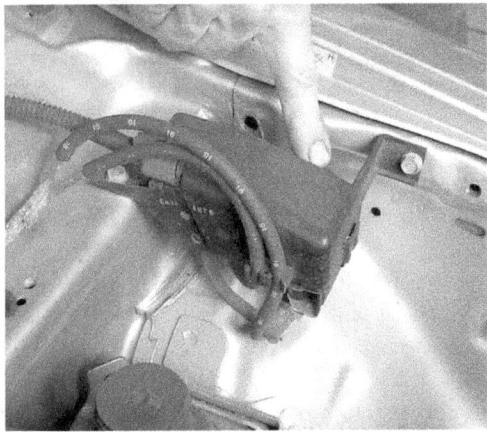

This 1996 Honda stored a P0401 (EGR low flow) DTC due to a defective vacuum control valve. But the bad valve doesn't store a DTC. Instead, the PCM identifies a low flow condition first, not a component failure. DTCs are sign posts.

# DTCs

## Professional Grade Diagnostics

Each section of this book is designed to prepare you for real world trouble-shooting. There are several ways to attack vehicle repairs, including:

- **Shotgunning** - This approach is simple: replace parts until the car is fixed. Inefficient and costly, this is the least desirable repair method.

- **Trouble Trees** - Trouble trees provide a guided diagnostic path to walk us through a repair process, step by step. Made of simple tests and "if-then" choices, the "tree" is commonly drawn on a paper as a series of text-filled boxes filled with directions, choices, and test values. To use the trouble tree, start at the top, follow the directions in the boxes, and arrive at a conclusion at the bottom. Ta Da!

  The final box—the repair recommendation—usually sounds something like: "Replace with known good widget and retest."

  This all sounds great, but trouble trees have real limitations. Sometimes they offer you two choices when neither is an option. Then what?

- **Professional Grade Diagnostics** - Skilled, experienced repair technicians commonly dislike trouble trees for the same reason Lance Armstrong would hate the thought of having training wheels installed on his bike. Too many limitations! Skilled technicians get answers faster by using DTC-defined pass-fail standards and testing carefully.

  If the DTC definition is poorly written and light on facts, a smart tech pans through the trouble tree, placer mining for test values; extracting diagnostic nuggets. He wants facts, not training wheels.

  Armed with this wealth of information, the tech attacks the problem scientifically, using traditional pinpoint tests performed with a digital multimeter, a fuel or vacuum gauge, a scan tool, etc.

  This diagnostic approach improves the speed, accuracy, and verification of the repair. It's the professional way to work.

  We recommend it.

## Putting DTCs to Work

• Look up the code definition. Don't limit yourself to the descriptor.

• Compare test standards in the DTC definition to actual vehicle conditions.

• Compare data values to test standards. Do test values match up to datastream and freeze frame?

• Use test criteria listed in the definition to simulate the conditions present when the DTC was stored. Professional application of defined test standards, under similar test conditions, is a fast and effective way to locate and repair failures that store DTCs.

• We hope you've gained a little respect for DTCs. Originally designed as an easily accessible diagnostic tool, the DTC shouldn't be something you rush past on your way to a more complicated solution.

## DTC Review

- DTCs are stored by the PCM when it detects a vehicle fault that may result in an unacceptable increase in vehicle emissions, from the tailpipe or from the vehicle fuel vapor containment system.

- DTCs are identified by a short label called a ***descriptor***.

- DTC "numbers" are *alphanumeric*, coded to indicate the main system, subsystem, components, and fault type for each DTC.

- DTC definitions should contain useful information about monitor test conditions and pass-fail standards.

- DTCs may be defined by the vehicle manufacturer or by the Society of Automotive Engineers (SAE).

- Some DTCs are stored in a single trip; others require two (or more) trips.

# THE DLC

**4**

# DLC

## Say AHHHHH!

The **Data Link Connector** is where you plug in the OBD II scan tool. We'll refer to it simply as the **DLC**.

The DLC has a standardized shape, with 16 cavities. It is designed to accept a cable connector from any OBD II-compliant scan tool interface.

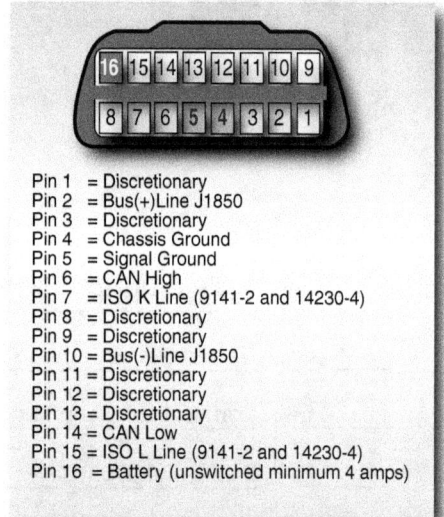

Pin 1 = Discretionary
Pin 2 = Bus(+)Line J1850
Pin 3 = Discretionary
Pin 4 = Chassis Ground
Pin 5 = Signal Ground
Pin 6 = CAN High
Pin 7 = ISO K Line (9141-2 and 14230-4)
Pin 8 = Discretionary
Pin 9 = Discretionary
Pin 10 = Bus(-)Line J1850
Pin 11 = Discretionary
Pin 12 = Discretionary
Pin 13 = Discretionary
Pin 14 = CAN Low
Pin 15 = ISO L Line (9141-2 and 14230-4)
Pin 16 = Battery (unswitched minimum 4 amps)

- Some DLC cavities are filled with electrical pins; others are empty.

- Some pin functions are mandatory; others are optional.

- Some pins are used for power and ground; others carry data.

- Optional pins are used at the discretion of the OEM (vehicle maker). Chrysler sometimes uses the CAN pins for a proprietary diagnostic interface called SCI.

For retrieving **OBD II generic data,** the standardized DLC shape means you don't need a box of specially shaped cable ends for your OBD scan tool interface. One size fits all.

(Enhanced data commonly requires special software and even special cables, but that is a different issue.)

### Generic Interface

Any OBD II-compliant cable end fits any OBD II DLC. In fact, the standardized DLC connector has created a whole new market in scan tools. Now anyone can purchase an affordable generic scan tool—or a laptop computer loaded with scan software—and connect it to the DLC.

The Auto Emissions Bible

## Generic Versus Enhanced Data

The **Generic OBD II interface** provides emissions-related data to a generic scan tool interface. Generic data include DTCs and Freeze Frame, datastream, and information about monitors and their status.

An **Enhanced** interface provides additional data, including information about non-emissions components. Scan tools from equipment makers who make dedicated scan tools for professional repair technicians normally include both **generic** and **enhanced** interfaces. Both are accessed through the DLC after the scan tool user selects the correct interface from a menu.

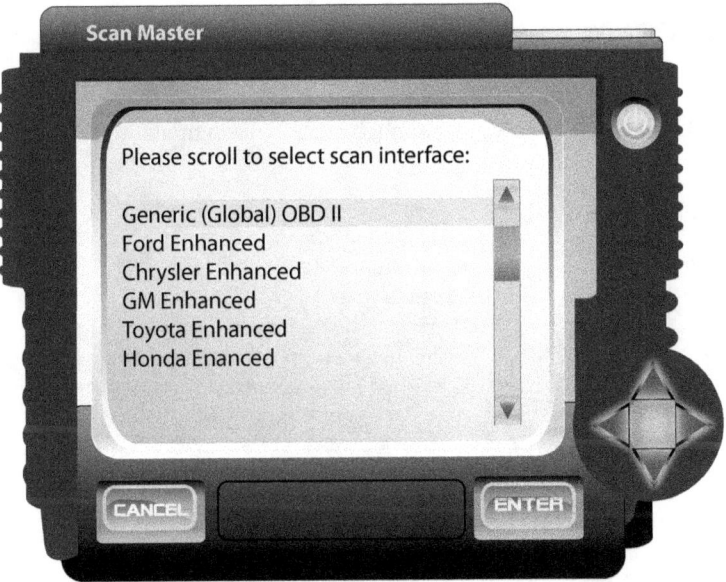

Dedicated scan tools are purpose-built to communicate with vehicles through multiple interfaces. Many professional grade scan tools can download and install software updates to keep them current. Our image shows a scan tool capable of communicating with several vehicle brands plus OBD II generic.

The scan tool interfaces at new car dealerships are a different story. OEM interfaces commonly have additional, proprietary capabilities that exceed features found in aftermarket scan tools.

# DLC

## Connect the Scan Tool

With the ignition switch turned to the OFF position, plug the scan tool cable connector into the DLC. Align the connector carefully as you push it in; don't try to force it into the DLC at an angle.

- Damaged pins at the scan tool cable end or DLC cause an assortment of data transfer problems, and may prevent communication altogether.

- Depending on the scan tool configuration, the cable connector may not have all 16 pins.

- Expect pins 4 and 5 (grounds) to extend outward farther than the rest; this is normal.

- The DLC shape is standard, although some pin assignments vary by make, model, and data protocol.

**Open Cavities**
Depending on the scan tool configuration, some cavities in the 16-pin DLC connector may be empty. Expect ground pins 4 and 5 to extend outward, slightly farther than the rest; this is normal.

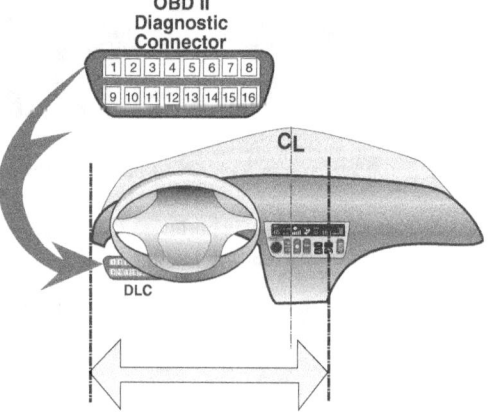

**Finding the DLC**
OBD II specifications indicate the general area for mounting the DLC. Ideally, you'll find it between the driver side door and right side of the center console.

Most of the time, you'll find the DLC beneath the lower edge of the dashboard.

## DLC Quick Tips

• The DLC shape is the same on all OBD II-compliant vehicles.

• The DLC has cavities for a maximum 16 electrical connections. Expect to see empty cavities in the DLC.

• The DLC should be located in the general area between the driver door and the right side of the center console. More recent standards narrow this zone, keeping the DLC on the driver side of the centerline, higher than the base of the steering column.

• The most common DLC location is beneath/behind the lower edge of the dash panel in the area near the steering column.

• Some DLCs are hidden in odd places or concealed by trim panels, ash trays, or carpeting. Some DLCs are located at the rear of the center console. (This is done by an engineer with too much time on his hands who wants to make the DLC more accessible to your rear seat passengers. We all know how much your mother-in-law loves to watch data on her scan tool on a long trip!) Later regulations eliminate this silliness.

• DLC damage may prevent data exchange between the scan tool and vehicle computer.

• A vehicle with a damaged or missing DLC cannot undergo a computerized scan tool emissions test.

The DLC shape is the same on all OBD II-compliant vehicles.

# DLC

## DLC Hiding Places

While the DLC is easily located and readily accessible beneath the dash in most cars, there are enough exceptions to annoy. Some OEMs have taken a perverse pleasure in showing how cute they can be when it comes to hiding the DLC.

A few "foreign cars" (whatever that means!) are repeat offenders: VW and BMW have sometimes built in a default layer of frustration with their patented snap-and-break plastic dashboard cover panels. Honda/Acura has been known to use what I call the Easter Egg Hunt approach, hiding the DLC behind the ashtray or mounting it on the right side of the center console behind a carpet flap. (CAUTION: Some of the dashboard trim/cover panels are *very expensive*.) When in doubt, consult a repair database, shop manual, or other information source to locate a hidden DLC.

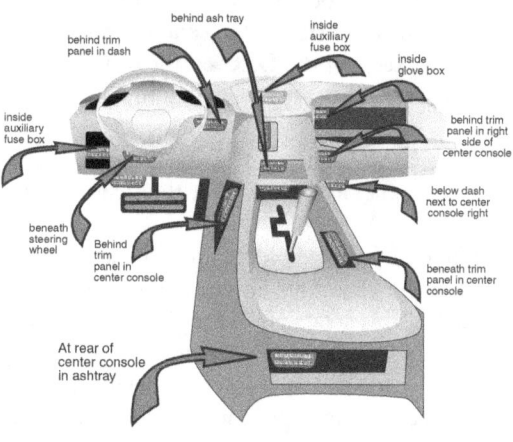

## DLC Damage

In areas where vehicle emissions are tested with a scan tool, the DLC *must* be accessible and undamaged, or the vehicle fails the test. If DLC damage prevents scan tool communication, the DLC must be repaired, and communications restored to allow testing.

When all else fails... read the directions! If the DLC isn't visible, take a good look at the underside of the steering column. This is where the vehicle manufacturer is supposed to put a sticker telling you just where the DLC is hidden. Be sure to check with your chiropractor before attempting this, or any other, under-dash gymnastics routine!

Use extreme caution when removing dashboard trim covers; some will not tolerate ham-fisted removal techniques. Some delicate plastic covers are painfully pricey.

Sneakier hiding places for the DLC include: behind ashtrays; beneath a carpet flap on the right side of the center console; and beneath fragile dashboard trim pieces. You may even find the DLC at the very rear of the center console, mounted there for ease of access by rear seat passengers who may want to plug in their scan tool for entertainment.

This one mounts sideways.

Is this upside down, or right side up?

It all depends on the vehicle and whether or not you are standing on your head at the moment.

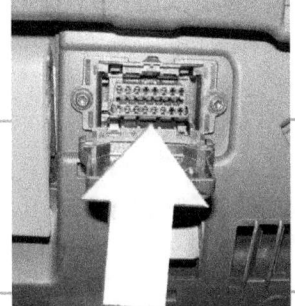

10 of the 16 cavities in this DLC contain electrical connections; the others are open.

# DLC

## Scan Tool Power

The DLC provides power and ground for scan tools that need external power. If a scan tool won't power up when you plug it in, check DLC pin 16 for battery voltage, and pins 4 and 5 for ground.

Some emissions test software tests DLC voltage values. A dead power or ground circuit may cause a motorist problems at an emission test.

**Note:** An increasing number of dedicated scan tools are self-powered by internal batteries, either conventional dry cell or rechargeable. Same goes for PC-based scan tools powered by a computer battery.

**CAUTION:** DLC pin 16 may be located in a fused circuit shared by other vehicle loads. If pin 16 is dead, check the fuses in the power distribution center (fuse panel).

**CAUTION: Never use the power and ground circuits at the DLC to operate vehicle accessories.** The DLC is *not* a power source for that new 400-watt amplifier!

## Failed Emissions Test?

If you cannot connect a scan tool to the DLC because it is damaged or inaccessible, the vehicle cannot be tested. Same thing happens when damaged DLC wiring or pins prevent data transmission: the vehicle is turned away from the test center until the DLC is repaired.

**Do NOT probe the DLC terminals from the front to test ground and voltage**.

Doing so may spread the connectors and damage them.

**Do NOT damage DLC terminals**

## DLC BOB

Now an admission. Having warned against probing the DLC with test probes, we need to confess that in the real world, probing from the backside is nearly impossible at many DLCs, and frustratingly awkward in the rest. If you must probe from the front of the DLC, use a *tiny* electrical probe and insert it with the utmost caution. Do not insert the probe and then let it hang with the weight of the test lead pulling it downward. The DLC connectors are very thin and easily distorted.

## Hi, BOB

There is a workaround. In fact, the method we recommend—especially if you need to test DLC circuits on a regular basis—is a DLC breakout box (BOB) test interface.

The BOB consists of a short wiring harnesses with male and female DLC connectors attached to opposite ends of a test console. Plug one end into the DLC and the other into your scan tool. Then you can safely probe at BOB's banana jacks to test circuits, while sitting comfortably upright. No more kneeling beneath the dash in an awkward position, probing by feel as you watch your voltmeter. Hit and mostly miss. You can even use your oscilloscope to monitor serial data pulses, if you want to go high tech.

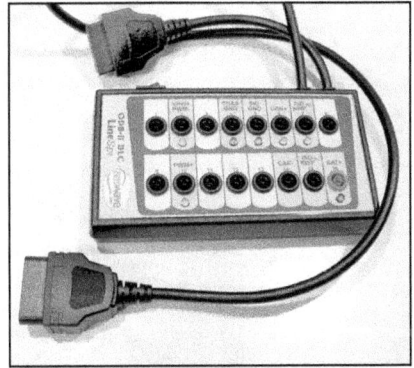

For those not familiar with the concept of a breakout box, a BOB is a patch harness with a test console. The console has banana plug test receptacles wired parallel to the main circuits, allowing you to test all circuits with your digital meter or lab scope, as they work.

A "smart" BOB like the one shown in our photo goes one step further. It has LEDs that light up or pulse to verify circuit condition, at a glance. Green means go and red indicates a problem in a circuit requiring additional tests. Duh.

This breakout box is available from two reputable sources: **Automotive Electronic Services** (*www.aeswave.com*) and **Automotive Test Solutions** (*www.automotivetestsolutions.com*).

# DLC

## Connection Cautions

We've been working on cars too long to take *anything* for granted. Sometimes the simplest mistake sets us up for failure, embarrassment—or both.

So be careful and please observe these DLC cautions:

- **Check your scan tool cable end for signs of wear or damage, especially bent pins.** Do so each time you connect to a DLC. Trying to cram a bent cable end connector pin into the DLC is a bad way to start a diagnosis.

- **Plug the cable end squarely into the DLC**; don't try to force it in at an angle. This is not always as easy as it sounds, especially if the DLC is mounted at an awkward angle in an inaccessible location.

- **CAUTION: Avoid kicking the connection.** Some DLC locations and extra long scan tool cable ends leave the cable hanging awkwardly beneath the dash, where it is easily kicked. Easy does it: a swift kick from a thick-soled work shoe can render the serial interface senseless. (Been there; done that.)

**One man's beef:**
Dear engineers: Some of you get wrapped up in minutiae and forget the obvious. Some DLCs are still placed on the left side, as shown in our photo. See how the interface cable hangs straight down? See how easy it is to kick?

Why not place the DLC somewhere safe, away from my size twelves? I am a repair technician, not a placekicker.

## Careful there, Sparky.

**Leave the ignition switch OFF until the scan tool cable is plugged securely into the DLC**: then turn the ignition switch to the ON position, or start the engine.

Truth be told, thousands of scan tools are connected and disconnected every day while the ignition is turned on or the engine is running, with no ill effects. But there are some vehicles that break out in a rash when this is done, and we don't want nasty letters saying we didn't warn you of the possibility.

The sidebar on this page illustrates one example of the kinds of problem that can be caused by leaving the ignition in the ON position while plugging in or disconnecting your scan tool.

• Some vehicle computers get stupid if the scan tool is disconnected with the ignition switch in the **ON** position: a few will even erase data or recalibrate. Some GM vehicles have been known to increase the odometer reading, and a few Chryslers changed their target charging voltage! For safety sake, we recommend that you turn the ignition switch to the OFF position before connecting or disconnecting a scan tool.

Here's an example:

**Tech Service Bulletin:** All 2003-2004 Mazda 6 and 2005 Mazda6 2.3L only.

**Description:** Some customer vehicles may have a false MIL illumination with multiple U-code DTCs. This condition may be caused if the WDS DLC cable is connected or disconnected while the ignition switch is in the ON position, or when the engine is idling.

**Important:** When connecting or disconnecting any scan tool equipment to the DLC-2 connector (under dash), the ignition switch must be in the **OFF** position; otherwise, the MIL may come ON and false CAN communication DTC codes may be set in the different CAN bus modules.

The following DTCs may be stored if this condition occurs: U1900-ABS; U2516-ABS; U1900-FF-IC; U2516-FF-IC; U0073-FF-PCM; U0073-FF-TCM; U0100-FF-TCM

It has been reported that some customer vehicles have failed the state emission OBD II test. This may be caused by the inspection station connecting or disconnecting the DLC cable when the ignition switch in the ON position, or when the engine is idling.

# DLC

## Texting the Vehicle

Texting is all the rage with young cell phone users, but OBD II computers have been doing it for years! It's hard to think of a car computer as hip or trendy, but OBD II was way ahead of its time!

The scan tool and PCM exchange information like two-thumbed typists. After you plug in the scan tool, the next step is to turn the ignition to the ON position or start the engine. At this signal, an OBD II scan tool and PCM exchange greetings, agree to use the correct data language, and establish a communication link.

An OBD II-compliant scan tool should check the data link and select the correct data protocol (language), automatically. We'll talk more about data protocols next, just remember that even the so-called "generic" interface may use one of several languages to communicate. (So much for standardization.)

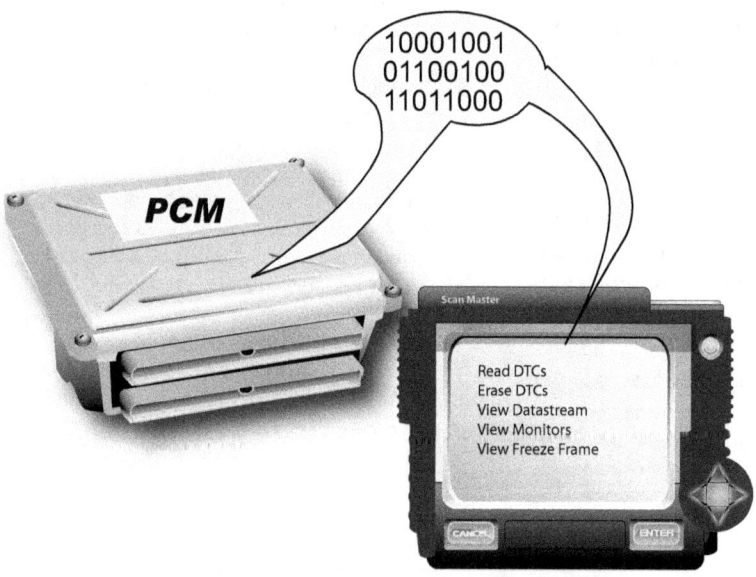

The PCM and scan tool exchange voltage pulses that are interpreted into meaningful data. The scan tool is the interface between machine and man, converting raw numbers into words and standard measurement values that humans can understand.

## Data Protocols

Just when we're picking up speed, we need to tap the brakes here and mention *data protocols*, an important issue for two major reasons:

- The scan tool must text messages to the PCM using the correct data protocol. Each protocol is a unique computer language. An OBD II-compliant scan tool should recognize the correct protocol and select it, automatically. (If I speak Spanish and you speak English, our conversation won't get far, will it?)

- Each data protocol has a maximum transmission speed, by design. First generation OBD II vehicles use one of four generic data transmission standards (some folks split the J1850 standard into three protocols, for a total of four). Why all these options were used or allowed is a mystery. It adds a needless layer of complexity, and none of the protocols is fast enough to get a speeding ticket in a school zone.

### Original OBD II Data Protocols

- **ISO 9141-2** - This protocol limps along at a very slow 10.4 kbps. Common in early Chryslers and many imports.

- **SAE J1850** - This protocol includes two subprotocols that operate at different pulse widths and transmit at different speeds: SAE J1850 Pulse Width Modulated (PWM) used by Ford (41.6 kbps), and SAE J1850 Variable Pulse Width (VPW) used by several OEMs (10.4 kbps).

> **Warning:** Some early OBD II data interfaces *still* cause "cuss and mumble" moments by refusing to talk to the first scan tool plugged into the DLC. The cure is often as simple as switching to a scan tool from another tool company. Many professional repair shops keep multiple scan tools handy for this reason. Thankfully, most generic communications problems have been resolved; just keep them in mind when working on last century OBD II vehicles.
>
> Some vehicles won't talk Key-On-Engine-Off; you'll have to start the engine.

- **ISO 14230-4** - (10.4 kbps) Keyword Protocol 2000, used by European OEMs. Similar to ISO 9141-2 but with enhanced diagnostic messages.

**Note:** No need to memorize all this. If your scan tool is working properly, it should automatically detect and use the vehicle protocol.

# DLC

## Protocols and Data Delivery Speed

Welcome to *datastream*. This topic is covered in greater detail in the next chapter. For now, however, you need to know that datastream is a line-item display of values for vehicle sensors and system parameters. It's a constantly updated list, all funneled to a single scan tool screen for our convenience.

**The Problem:** Sucking all the available datastream values through old, generic protocols is slow going; like draining a swimming pool through a drinking straw. A slow generic protocol is a data traffic bottleneck.

Things are slowest when you want to view **all** available datastream items on a single screen: the interface must download all parameters, one at a time, and then return to the top of the data list and start again. In the modern world of computers, a generic data rate of 10.4 kbps (kilobytes/second) is the equivalent of sending smoke signals.

This may not matter for temperature sensor parameters that change slowly. But a slow datastream protocol can't update fast-changing signals like rpm and $O_2$ sensor voltages in real time. Not even close. In extreme cases, you may think the data link is down—or that the scan tool is defective.

The cure is more of an crude workaround than a real solution: For faster data updates, select fewer PIDs. Fewer PIDS mean less data and that means faster updates of selected PID values.

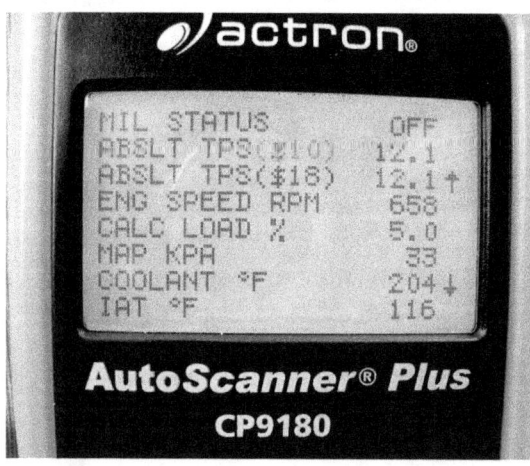

Each of these data parameters is updated—one at a time—as data arrive from the PCM.

When the entire list is updated (including the PIDs that won't fit on the screen), the process begins again. If data transfer is slow—and many parameters are selected—the values will refresh on the screen very s-l-o-w-l-y.

## Controller Area Network

The current data protocol, Controller Area Network (CAN) supersedes all previous protocols.

This is good news, for several reasons.

• While CAN is available in several speeds, it is always faster than its pre-decessors. That means faster scan tool screen updates.

• The CAN interface eliminates most scan tool-to-PCM connection problems caused by protocol recognition errors.

• CAN vehicles support a larger number of data parameters. This gives us access to a longer datastream list with more repair data. (Compare the short data list from early OBD II on page 81 to the CAN PID list on pages 83-86.)

• CAN handles data more efficiently in networks where messages must be sorted by priority.

Some scan tools do not support CAN. Others offer an upgrade allowing you to add the protocol. If you still have your golden oldie scanner from the nineties, it may or may not have been upgraded to speak CAN. If that's the case, you'll need to upgrade your old scan tool (assuming that is possible) or purchase a new, CAN-compatible scan interface.

### Which Cars Have CAN?

If you just pulled out of the showroom in your shiny new car, rest assured it's CAN-compliant. CAN has been phased in by car makers over the last several years. All 2008 and newer vehicles sold here in the good old US of A are CAN-standard.

## DLC Review

- Scan tool communication occurs over a serial interface where data is sent in both directions by serial voltage pulses.

- Early interfaces may have programming issues that prevent communication with some scan tools.

- The scan tool should automatically detect the correct data protocol and allow communications.

- If your scan tool will not communicate with a vehicle, it pays to try a different scan tool before resorting to more complicated diagnostics. In some cases, the engine must be running for communication to occur.

- Generic scan interfaces before CAN are slow. Controller Area Network (CAN) interfaces are much faster than the original OBD II protocols and provide more data.

- To prevent unwanted problems, connect and disconnect the scan tool only when the ignition is off.

# DATASTREAM 5

# Datastream

## Spies Everywhere!

*Datastream* is exactly what its name suggests; it's a stream of data carried over a wire by voltage pulses. Pulse sequence, amplitude, and duration can be arranged into patterns that carry information. An analogy to Morse Code should work here, assuming you didn't sleep through American History in middle school.

Using a scan tool plugged into the standard OBD II data port, a carbon based life form can text the vehicle computer and ask for information. If all goes according to plan, the PCM gets the message and texts back the desired data. (Heck, you've seen Mr. Scott do this on *Star Trek* to make sure the warp engines are working.)

## Small Things Make Us Happy

If you ask it to do so, the PCM will stream data that contains information about the exact status of various emissions-related vehicle parameters.

This may seem like a ho-hum event until you realize that OBD II broke some major ground by allowing a data exchange over a standardized interface. In 1996, when there was absolutely *nothing* standard about scan tool interfaces, this was a bigfriggin deal. At that time, individual OEM interfaces were commonly limited and totally nonstandard.

Then came OBD II. All at once—for the first time ever—*anyone* could connect a *standard* scan tool to a *standardized* data connector, and look at a *standard* set of data parameters. The original "generic" data list for emissions was admittedly a short one, but a welcome improvement, nonetheless.

## What's a PID?

A simple datastream display should get us off on the right track. Here's a list of early generic vehicle data parameters.

| | | | |
|---|---|---|---|
| Calculated Load | LOAD_PCT | | % |
| ✓ Engine Coolant Temperature | ECT | 93.0000 | Deg C |
| ✓ Short Term Fuel Trim Bank 1 | SHRTFT1 | 3.1250 | % |
| ✓ Long Term Fuel Trim Bank 1 | LONGFT1 | 2.3437 | % |
| ✓ Intake Manifold Absolute Pressure | MAP | 28.0000 | kPa |
| ✓ Engine RPM | RPM | 702.2500 | RPM |
| ✓ Vehicle Speed Sensor | VSS | 0.0000 | km/h |
| ✓ Ignition Timing Advance for #1 Cylinder | SPARKADV | 6.5000 | deg |
| ✓ Intake Air Temperature | IAT | 31.0000 | Deg C |
| ✓ Absolute Throttle Position | TP | 9.4118 | % |
| ✓ O2 Bank 1 - Sensor 1 | O2B1S1 | 0.8150 | V |
| ✓ FT Bank 1 - Sensor 1 | FTB1S1 | 2.3437 | % |
| ✓ O2 Bank 1 - Sensor 2 | O2B1S2 | 0.2400 | V |
| ✓ FT Bank 1 - Sensor 2 | FTB1S2 | 99.2187 | % |

Right now, these may not mean a lot to you, but they will. Each item in this list is a parameter, and each parameter stands for a component or system that reports to the vehicle computer.

- Parameters have labels and a shorthand reference. For example, the **Engine Coolant Temperature** sensor is identified by name and also by its three-letter abbreviation: **ECT.**

- **P**arameters and their **ID**entification labels are commonly referred to as **PID**s. You'll be seeing this term a lot from here on out.

- Each PID has its own value. PIDs are displayed as a voltage value or as a calculated value, in units indicating position, temperature, pressure, speed, etc. These values are listed in a column to the right.

In a repair shop, you'll hear technicians refer to the "ECT PID" or the "O2 sensor PID" when they discuss line-items in datastream.

If one of the values listed next to a PID looks funky, you may hear a technician say something like, "That O2 sensor PID is not changing." Or, "That O2 sensor PID is too low."

# Datastream

## Data Display Samples

Here are sample datastream displays on different scan tool screens. Each of these is a dedicated scan tool. Screens are pretty small compared to PC-based scan tools with large monitors.

This scan tool from Actron lists all PIDs, with room for 8 separate items per screen. The arrows to the right of the screen indicate that additional PIDs are available by scrolling.

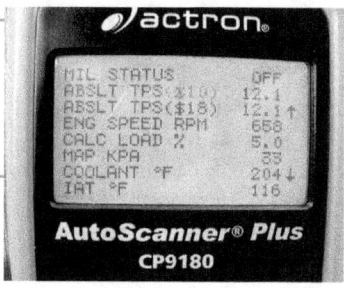

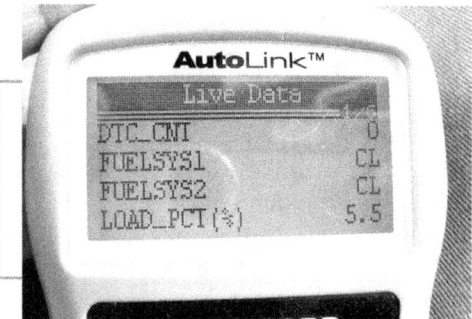

This inexpensive scan tool from AutoLink costs less than a hundred bucks. It displays a maximum of four PIDs per screen, so expect to scroll to view all PIDs.

This professional grade scan tool displays nine PIDs per screen. Dedicated scan tools like this have limited display "real estate," especially when compared to PC-based scan tools.

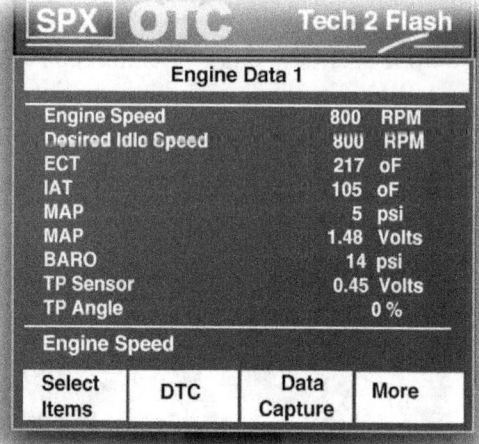

# Datastream

## Original OBD II Generic Datastream

The MIL is on. You have retrieved DTCs. Now it's time to look at generic datastream to see if the conditions that stored the DTC are still present.

OBD II generic datastream displays a select number of emissions-related data parameters that have a direct and significant effect on engine performance, fuel control, and emissions.

### Generic PID List

Look at the items in the generic PID list on this page. These PIDs may be viewed in datastream on a generic scan tool.

The items shown here are the original "short list" that came with pre-CAN OBD II vehicles. No frills here, folks.

| | |
|---|---|
| Calculated Load | percent |
| Engine Speed | rpm |
| Absolute Throttle Position | percent |
| Vehicle Speed | km/h-mph |
| MIL Status | on/off |
| Fuel Control System Status | closed/open |
| Fuel Trim (STFT and LTFT - B1, B2) | percent |
| Secondary Air Status | on/off |
| Ignition Timing Advance | degrees/rotation |
| Intake Air Temp | degrees/temp |
| Engine Coolant Temp | degrees/temp |
| MAF | gm/sec -lb/min |
| MAP | kPa-inHg |
| Oxygen Sensor Output | volts |
| PTO Status | active/inactive |

Here's the original list of pre-CAN, generic PIDs.

The information here is all powertrain related, and engine specific. Note that there is no information given about the vehicle transmission or evaporative emission system. Too bad, because either one can store a DTC and turn on the MIL.

Note: A fuel pressure PID is listed in the original OBD II specification, although we've never seen it used in those vehicles. If you see it in a real pre-CAN vehicle, please drop us a line!

If you want serial data or test values for the tranny, evaporative emissions system, or any other part of the vehicle, you won't find it here. You'll need to switch to an OEM enhanced interface for non-OBD II information like that.

**Long story short:** The original OBD II datastream list is short. It provides limited information, most of it about components and systems related to engine performance.

# Datastream

## Generic Data Interface

The generic OBD II data list exists to provide basic information about emissions-related systems and components. In the original OBD II standards (before CAN), generic datastream items are limited to powertrain components.

• **The generic OBD II interface communicates with the PCM using one of several data protocols.** A protocol is nothing more than a computer language. All original OBD II protocols are slow. They were slow in 1996 and seem even slower now when we compare them to the CAN (Controller Area Network) protocol that replaces them all. Slow protocols mean slow data transfer and slow scan tool screen updates.

Slow scan data updates are a real pain when you want to see data values updated in real time–or something close to it. Datastream updates using the original protocols are so slow under some conditions, that it's like reading the weather report from yesterday's newspaper.

• **All data reporting in generic datastream indicates the *actual* value of the sensor input, as seen by the PCM.** Please read that last sentence again. This is extremely important. Modern vehicle computers can substitute a sensor value to keep the vehicle running. It may not be the best value, but it keeps you from calling a cab and a tow truck.

True sensor value is a big diagnostic clue if a smarty-pants PCM plugs in a substitute value in place of the missing signal. Some PCMs will lie to you with a straight face. If you are viewing the enhanced factory datastream, the "plug in" value for a missing input may look so normal in datastream that you won't question it.

(See pages 88-91 for examples.)

# Datastream

## CAN Data PIDS - Honda Fit

Here's a look at a data screen from a 2011 Honda Fit. Yikes Look at how much more data we get from this vehicle data list than we did from first generation OBD II generic vehicles.

This is a huge improvement. Compare it to the list on page 81.

| Supported PIDs | Abbrev | Data | Units |
|---|---|---|---|
| ✓ Calculated Load | LOAD_PCT | 21.1765 | % |
| ✓ Engine Coolant Temperature | ECT | 201.2000 | Deg F |
| ✓ Short Term Fuel Trim Bank 1 | SHRTFT1 | 5.4687 | % |
| ✓ Long Term Fuel Trim Bank 1 | LONGFT1 | 3.9062 | % |
| ✓ Intake Manifold Absolute Pressure | MAP | 7.6778 | HG |
| ✓ Engine RPM | RPM | 691.0000 | RPM |
| ✓ Vehicle Speed Sensor | VSS | 0.0000 | mph |
| ✓ Ignition Timing Advance for #1 Cylinder | SPARKADV | 4.5000 | deg |
| ✓ Intake Air Temperature | IAT | 129.2000 | Deg F |
| ✓ Air Flow Rate from Mass Air Flow Sensor | MAF_g/s | 1.6400 | g/s |
| ✓ Air Flow Rate from Mass Air Flow Sensor | MAF_lb/m | 0.2165 | lb/m |
| ✓ Absolute Throttle Position | TP | 12.5490 | % |
| ✓ O2 Bank 1 - Sensor 2 | O2B1S2 | 0.7100 | V |
| ✓ O2 Bank 1 - Sensor 2 | FTB1S2 | 5.4687 | % |
| ✓ Time Since Engine Start | RUNTIME | 342.0000 | s |
| ✓ Distance Traveled While MIL is Activated | MIL_DIST | 0.0000 | miles |
| ✓ Commanded EGR | EGR_PCT | 0.0000 | % |
| ✓ EGR Error | EGR_ERR | 99.2187 | % |
| ✓ Commanded Evaporative Purge | EVAP_PCT | 29.0196 | % |
| ✓ Fuel Level Input | FLI | 41.5686 | % |
| ✓ Number of Warm-ups Since DTCs Cleared | WARM_UPS | 60.0000 | |
| ✓ Distance Since Diagnostic Trouble Codes Cleared | CLR_DIST | 956.3400 | miles |
| ✓ Evap System Vapor Pressure | EVAP_VP | 385.5000 | Pa |
| ✓ Barometric Pressure | BARO | 28.9394 | HG |
| ✓ Bank 1 - Sensor 1 (Wide Range O2S) (mA) | WRERB1S1 | 1.0085 | Ratio |
| ✓ Bank 1 - Sensor 1 (Wide Range O2S) (mA) | WRO2B1S1 | -0.0312 | mA |
| ✓ Catalyst Temperature Bank 1, Sensor 1 | CATEMP11 | 407.1200 | Deg F |
| ✓ Control Module Voltage | VPWR | 14.2500 | V |
| ✓ Absolute Load Value | LOAD_ABS | 16.0784 | % |
| ✓ Commanded Equivalence Ratio | EQ_RAT | 1.0043 | Ratio |
| ✓ Relative Throttle Position | TP_R | 1.1765 | % |
| ✓ Absolute Throttle Position B | TP_B | 30.1961 | % |
| ✓ Accelerator Pedal Position D | APP_D | 19.6078 | % |
| ✓ Accelerator Pedal Position E | APP_E | 9.8039 | % |

On the next two pages, we give a brief explanation of OBD II CAN generic PIDs.

# Datastream

## CAN Generic PIDs

**Absolute Load** - A value indicating air mass taken in per intake stroke, expressed as a percentage.

**Absolute Throttle Position** - Actual throttle position expressed as a percentage, in a range from zero to 100%. The number increases with increased pedal movement.

**Accelerator Pedal Position** - True accelerator pedal position, expressed as a percentage, in a range from zero to100%. The number increases with increased throttle opening.

**Air Flow Rate MAF** - The mass of air taken into the engine, measured by the mass air flow sensor.

**Ambient Air Temperature** - Temperature of the air surrounding the vehicle.

**Barometric Pressure** - Pressure of the air surrounding the vehicle (kPa or inHg).

**Calculated Load Value** - Indicates the percentage of engine capacity being used. The value reaches 100% at wide open throttle.

**Catalyst Temp Bank X** - The temperature of the catalyst substrate. This is an inferred value.

**Commanded Equivalence Ratio** - A value used to calculate the target A/F ratio being commanded by the PCM. Like lambda, 1.0=14.7 :1. Numbers lower than 1.0 indicate a "go rich" command. Those above 1.0 indicate a "go lean" command from the PCM.

**Commanded EGR** - The PCM command to open the EGR, expressed as a percentage, in a range from zero to 100%, with 100% indicating a PCM command for maximum EGR flow.

**Commanded Evaporative Purge** - The PCM purge command, expressed as a percentage, in a range from zero to 100%, with 100% indicating a PCM command for maximum purge flow.

**Commanded Throttle Actuator** - The PCM throttle opening command, expressed as a percentage, in a range from zero to 100%, with 100% indicating a PCM command for maximum throttle opening.

**Control Module Voltage** - PCM power, in volts DC.

**Distance Traveled While MIL On** - Miles/km driven since the MIL was commanded on.

**Distance Since DTCs Cleared** - Miles/km driven since DTCs were erased.

**EGR Error** - The difference between EGR commanded and actual states.

**Engine Coolant Temperature (ECT)** - The temperature of the engine coolant, in degrees.

# Datastream

## CAN Generic PIDs

**Engine RPM** - Engine speed in crankshaft revolutions per minute.

**EVAP System Vapor Pressure** - Fuel system (EVAP) vapor pressure measured by a transducer inside the fuel system, commonly inside the fuel tank or attached to the EVAP canister.

**Fuel Level** - The fuel level inside the fuel tank, expressed as a percentage.

**Fuel Rail Pressure (gauge)** - The fuel rail gauge pressure (not absolute pressure).

**IAT** - Intake Air Temperature.

**Ignition Timing Advance** - Ignition timing advance for #1 cylinder (not including mechanical advance).

**Intake Manifold Pressure** - Manifold Absolute Pressure (MAP), measured in (kPa or inHg).

**Long Term Fuel Trim** - The learned, long term PCM fuel correction, reflecting long term fuel delivery trends.

**Minutes Run with MIL On** - The accumulated time the engine has run with the MIL on.

**O2 Sensor Voltage BX-SX** - The voltage for conventional oxygen sensors that operate in a 0-1 volt range, identifying the sensor by bank (B1, B2) and location in the exhaust (S1, S2).

**O2 Sensor BX-SX Wide Range mA** - The milliamperage for linear or wide-ratio oxygen sensors.

**O2 Sensor BX-SX Wide Range V** - Voltage for linear (wide-range) oxygen sensors.

**Relative Throttle Position** - A learned throttle position value.

**Short Term Fuel Trim-BX (1-2)** - PCM fuel adjustments required to maintain a stoichiometric mixture in closed loop.

**Time Since DTCs Cleared** - The length of time since DTCs were cleared with a scan tool.

**Time Since Engine Start** - The length of time since the engine was started.

**Vehicle Speed** - Vehicle road speed (distance/time)

**Warm-ups Since DTCs Cleared** - The number of warm-up cycles recorded since all DTCs were cleared with a scan tool.

# Datastream

## The Value of Generic Datastream

The improved and enlarged CAN data list on the preceding pages is a *huge* improvement over the brief, original parameter list introduced back in 1996.

- **We get more useful more data about code storage and erasure.** We are shown engine run time, how long the MIL has been illuminated, in time and miles driven, and how long ago DTCs were erased. This information is critical when we are trying to determine how often a problem occurs, especially an intermittent one.

- **We also get a new PID called Commanded Equivalence** that indicates the PCM air/fuel ratio target, a much more precise indication of closed loop fuel control than mere fuel trim. Now you can compare the lambda command (Commanded Equivalence Ratio) in datastream to the true lambda value on your exhaust analyzer.

- **Several emissions subsystems are newly represented.** EGR gets a command PID plus an error PID to indicate if the command from the PCM results in the proper EGR response.

- **The evaporative system, ignored by early OBD parameters, is now represented.** We can expect to see important data about fuel level, system pressure, and purge command.

- **The air/fuel sensor now has its own PIDs, and properly displays voltage and milliamperage,** something that was not possible under the original OBD specs; AF sensors were not in use when they were written.

- **A wonderful new PID lets us monitor control module voltage.** This eliminates a potential performance issue without the cost in time and aggravation required to perform a loaded pinpoint voltage test at the PCM.

- **New PIDs have been added to monitor by-wire throttle systems,** a new technology standard that has all but eliminated the throttle cable.

Bottom line? Improvements in CAN datastream allow a skilled technician to triage a patient vehicle and assess its vital signs in minutes. We should all thank the parties responsible for this huge improvement, and learn to maximize the benefits of these added PIDs through regular practice.

# Datastream

## Data Traps

In general, scan tool interfaces connected through the OBD data link connector (DLC) allow us to sample two general classes of data: OBD II **generic** data, dedicated to emissions, and a separate, **manufacturer-specific** data class that we'll call **enhanced** data. Enhanced datastream commonly displays additional PIDs including non-emissions items.

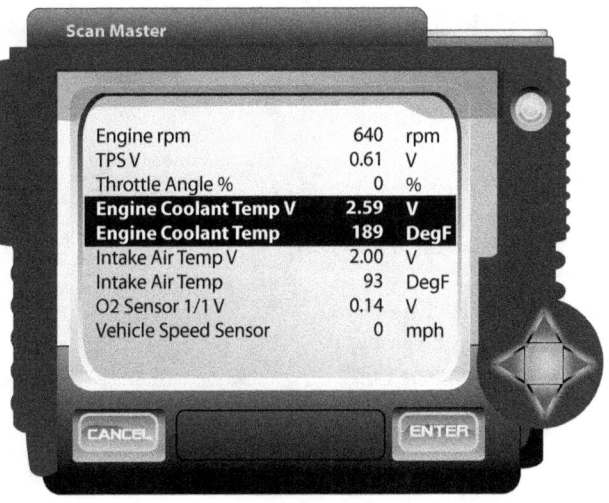

Core generic PIDs commonly number between 20 and 40. Manufacturer-specific PIDs displayed in the enhanced interface may number in the *hundreds.*

While there is some overlap in the PID lists of the two data classes, there is a big difference in the **way** the data are represented.

Let's begin with the enhanced scan data display shown above. This display is typical of what you see with the "factory" interface selected on your scan tool. (There are many more PIDs than will fit on the screen.) Concentrate on the Engine Coolant Temp sensor (ECT) PIDs; there are two, and we highlighted them on a black background. One displays ECT voltage, and the other, an "interpretation" of that voltage, indicating coolant temperature, in degrees.

Since the normal operating range for the ECT is between zero and 5 volts, the 2.59 volt sensor value and the interpreted temperature of 189°F look logical. These readings are normal. They should be, because this car has no DTCs. It is good to go.

Now let's see what the scan tool displays in both generic and enhanced data when there is an open in the ECT circuit. This is where we have to pay special attention to both the interface and the data, or get sidetracked.

# Datastream

## Substituted Values

Here is the same vehicle from the previous page. This time, however, it has a problem. The MIL is on and a DTC is stored: **P0118 - Engine Coolant Temperature Sensor High Input.**

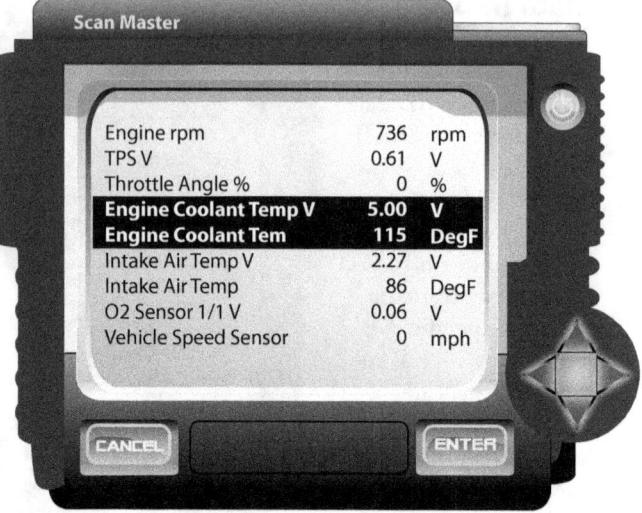

| | | |
|---|---|---|
| Engine rpm | 736 | rpm |
| TPS V | 0.61 | V |
| Throttle Angle % | 0 | % |
| **Engine Coolant Temp V** | **5.00** | **V** |
| **Engine Coolant Tem** | **115** | **DegF** |
| Intake Air Temp V | 2.27 | V |
| Intake Air Temp | 86 | DegF |
| O2 Sensor 1/1 V | 0.06 | V |
| Vehicle Speed Sensor | 0 | mph |

If we glance at data using the **enhanced interface**, we see an ECT temperature of 115 °F, and may

The DTC tells us ECT sensor voltage is too high. Scan data displayed in the enhanced interface confirm it.

wonder what all the fuss is about. That is a possible value and maybe even a reasonable coolant temperature under some circumstances. Besides, if the ECT sensor circuit were truly open, we would see a very low temperature, commonly below freezing—or even below zero degrees F! Wouldn't we?

We need to look at all data before jumping to a bad conclusion. **The 5.0 volt sensor voltage displayed in the data line directly above the calculated temperature is a dead giveaway that the circuit is open.** It is full reference voltage. It's what we would see if the sensor were unplugged. This circuit is open!

So where did the 115 °F come from? It is a **plug-in value**, substituted by the computer to keep the car running. **It is a lie.**

**To spot plug-in values, always compare the interpreted value to the signal voltage to be sure they BOTH make sense.**

**Next:** Same car, same problem when viewed in the generic interface.

# Datastream

## Start With Generic

Same car. Same DTC. But this display is different, isn't it?

This time, our scan tool is displaying OBD II *generic* data.

Notice anything different? Yeah, that temperature value looks like it's coming from Point Barrow Alaska, in February. Minus 40 degrees F.

Write this down someplace where you won't lose it:

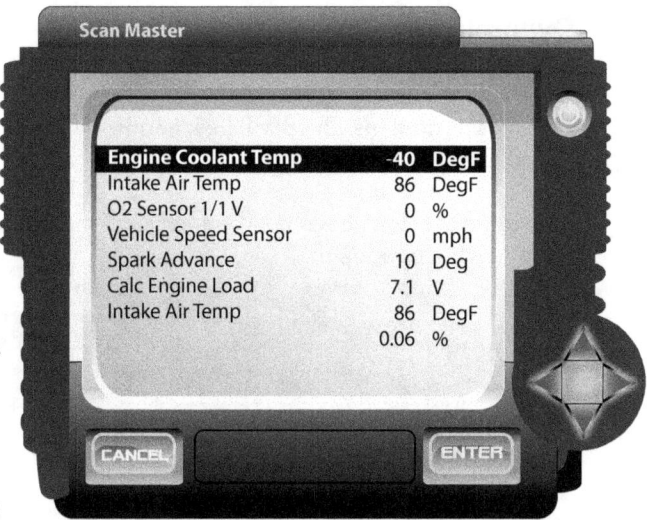

| Scan Master | | |
| --- | --- | --- |
| **Engine Coolant Temp** | **-40** | **DegF** |
| Intake Air Temp | 86 | DegF |
| O2 Sensor 1/1 V | 0 | % |
| Vehicle Speed Sensor | 0 | mph |
| Spark Advance | 10 | Deg |
| Calc Engine Load | 7.1 | V |
| Intake Air Temp | 86 | DegF |
| | 0.06 | % |

The -40 degree value shown here is a clue that the ECT is open. The 5 volt ECT signal voltage we saw in enhanced data on the previous page confirms it.

 **OBD II generic interface displays the *actual* calculated value in both datastream and freeze frame.**
**You will see no substitutions. No lies.**

**Compare generic and enhanced data, when both interfaces are available.** Combining the -40 degree reading from generic interface with the 5.0 volt reading from the enhanced interface assures us that we do indeed have an open circuit.

After the repair? We'll double back to the scene of the crime and review the same data set again. Only this time data ought have the same logical characteristics we saw in our first example from a " good car."

Don't be fooled by a substituted value.

# Datastream

## Dynamic Substitution

The plug-in value won't always be a static value. The PCM won't just default to a pre-programmed number and stick with it. Sometimes, the PCM makes *dynamic* substitutions. These will look normal on the scan tool screen, unless you are careful.

The scan tool below shows how one vehicle computer compensates for a failed MAP sensor input.

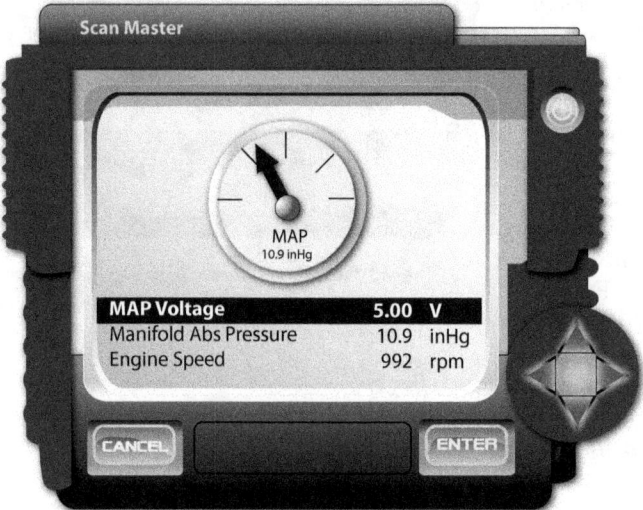

The data displayed in our image is gathered using the OEM enhanced interface, not OBD II generic. The scan tool combines line-item data PIDs with a simulated analog vacuum/pressure gauge, complete with needle.

The MAP value is consistent with what we might see from an engine running at this speed. If we are not careful, we'll accept this realistic substitute as an acceptable value. A closer look shows that the MAP signal voltage is pegged at full reference voltage: 5 volts. This is what we would see with an open-circuited MAP signal input to the PCM.

Where does the 10.9 inHg come from? The PCM is calculating what MAP *should* be at this particular rpm, and then plugging that *calculated* value into datastream.

## Do not hurry through datastream and freeze frame values. Study them as a surgeon studies x-rays, looking for clues!

# Datastream

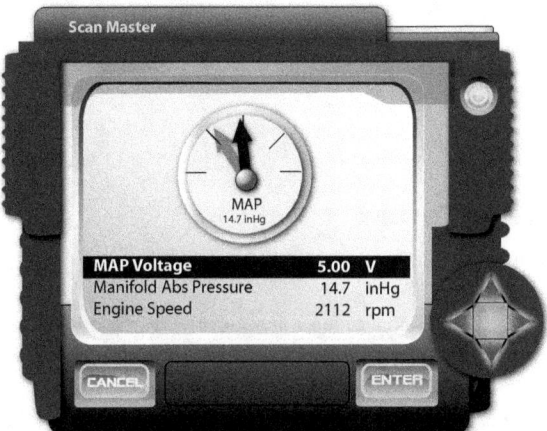

To make this even harder to spot, MAP PID values in this vehicle **change** as we raise the engine speed from 992 to 2112 rpm. It looks like the MAP is responding normally as the throttle opens and manifold pressure increases. That's why the 5.0 volt MAP value is an essential clue in this little mystery. **It never changes!**

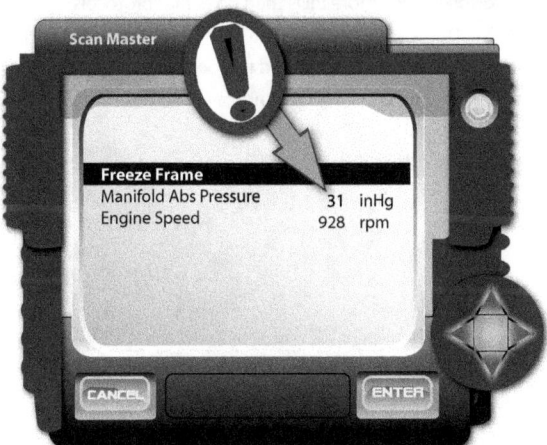

Let's cross reference live data with Freeze Frame to prove our point. Remember, Freeze Frame and Datastream in Generic OBD II **do not show substituted values** plugged in by PCM substitutions; only raw data.

The Freeze Frame MAP PID displays the maximum MAP sensor value available in this system: 31 inHg. This maxed-out value plus the 5 volt signal voltage from live datastream confirm that there is an open in the MAP sensor or its circuit.

## Datastream Review

- Datastream is a stream of data carried by electrical pulses.

- A PID is a **P**arameter **ID**entification, a line item in a list of components and system conditions, and their values.

- The number of PIDs displayed at one time on a scan tool screen varies. To see all PIDs, scrolling is necessary in some scan tool interfaces.

- Scan tools and vehicle computers communicate using different languages called ***protocols***.

- CAN vehicles display many more PIDs than do earlier protocols.

- Powertrain computers may calculate and use both pre-programmed or calculated substitutions when critical sensor inputs are lost.

- OBD II generic datastream does not display substituted values.

# MONITORS

# Monitors

## Monitors May Be Murky

If you are a first time scan tool user, diagnostic trouble codes (DTCs) and live datastream will commonly be your first stops as you navigate through various screens and menus. Then you'll tackle Freeze Frame, and things move along well as you add to your diagnostic skill set.

Then we get to a speed bump called Monitors. Monitors are a little tougher to grasp, at least initially. In fact, for the OBD II beginner, monitors can be downright puzzling. It's easy enough to say what monitors are for: to test the vehicle. It's a lot harder to describe how they go about their business and how they work—or don't work—as is sometimes the case.

Monitors are complex software programs; subroutines inside the larger OBD strategy. Many of these subroutines have subroutines of their own. Their overall intent is simple: to test the vehicle. Their operation, however, can be a mystery at times. Monitors can stress you faster than a teething toddler.

Don't feel bad if the concept of monitor strategies takes some study, friend, because auto repair professionals have had to go through a bit of a learning curve on this subject. Even experienced technicians can still be heard growling when a balky monitor refuses to run as it should.

Let's start by defining a monitor and listing some monitor features and characteristics, and then move on to the fine points.

# Monitors

## What Is a Monitor?

Maybe the best way to understand what monitors **are** is by asking what monitors **do**. On the surface, it's simple: they run tests. And the results of those tests determine whether the MIL comes on—or stays off.

For a working definition, lets go with: *Monitors are software programs that test different parts of the vehicle.*

## How Many Monitors Are There?

Monitors are specialists, testing vehicle subsystems and their components.

• Each OBD II vehicle has multiple monitors—all of them different.

• Monitors must run in a logical sequence, since some monitors need test information that's available only in test data returned by other monitors.

• Monitors must be scheduled to run in a specified order or they may interfere with one another.

• Monitors run only when conditions are right for them to run. Conditions for running a monitor are carefully defined. The reason? If monitors run under adverse conditions, they are more likely to store false codes.

• If a monitor does not run, it cannot store a DTC.

This is all very important. But it's just the start. There's more to know about monitors, especially if you need to pass an OBD II scan tool emissions test. (For more, see **Chapter 11, Passing the Emissions Test.**

# Monitors

## Enabling Criteria

We just mentioned that vehicle operating conditions must be "just right" before a monitor will run. An example may help. Let's say you want to test a new sun block lotion. But it's raining today. Conditions are not right to test sun block, so you'll wait for the sun to come out, and then try again.

To keep mere mortals like you and me guessing, engineers branded the set of conditions needed to run a monitor, *enabling criteria*. Calling them "proper test conditions" or even "suitable test conditions" would be easier to understand, but plain talk gets scarce sometimes in the automotive world.

Another working definition: ***Enabling Criteria* is just a fancy term for the conditions needed to run a monitor.**

 The chart below displays enabling criteria for one monitor. Keep them in mind, because we'll return to this subject later in this section when we try to "run monitors" so we can pass an emissions test

| Fault | DTC | DTC Definition | Test Values |
|---|---|---|---|
| O2S Heater Circuit Malfunction Bank 1 | P0141 | PCM monitors time to activity following a startup to determine if the heater is operating. | Elapsed time for the $O_2$ sensor to reach 150 millivolts above or below bias voltage. |

| Enable Conditions | Test Time/Frequency | MIL Test Type (A or B) | Special Notes |
|---|---|---|---|
| No Misfire DTCs<br>No Crank Sensor DTCs<br>No Injector DTCs<br>No MAF DTCs<br>No MAP DTCs<br>No TP DTCs<br>No EVAP DTCs<br>No IAT DTCs<br>No Fuel Trim DTCs<br>ECT<35°C<br>ΔECT-IAT≤6°C<br>Avg MAF<24g/s | From cold start to a maximum time of 450 seconds | Continuous<br>Type B | > = greater than<br>< = less than |

These are the enabling criteria for testing an oxygen sensor heater in a specific vehicle. (Enabling criteria for the same monitor are probably different in another vehicle.) These are the vehicle operating conditions required before the test will run in this vehicle, plus the pass-fail test standards. (Remember the importance of code definitions?)

Information like this eliminates guesswork, helps us verify our repairs, and lets us run monitors in less time.

## Monitors Come In Two Flavors

There are two kinds of monitors: continuous and non-continuous.

### Continuous Monitors

Continuous monitors run, well...*continuously*. Continuous monitors keep testing whenever the vehicle is in operation—whenever vehicle conditions are just right—when **enabling criteria** are satisfied.

If they are not? Continuous monitors may be **suspended** if operating conditions might generate a false DTC. Examples: A low fuel level might cause a misfire; an extremely full fuel tank might affect the fuel system vapor leak test. Ambient conditions and some kinds of vehicle operation can suspend one or more of the onboard monitors.

OBD II vehicles all have the same three **continuous** monitors: **comprehensive component, fuel, and misfire**. These three *amigos* are really important because they test the core components and systems responsible for combustion efficiency. Cleaner combustion is the goal, since it reduces the amount of toxic exhaust that must be cleaned by the catalyst.

If you look at vehicle monitors on your scan tool, **you'll always see these same three continuous monitors**. To confuse the novice scan tool user, continuous monitors will always be listed as complete...or done...or YES... depending on the scan tool. Their readiness status never changes (making us wonder why their status is listed in the first place).

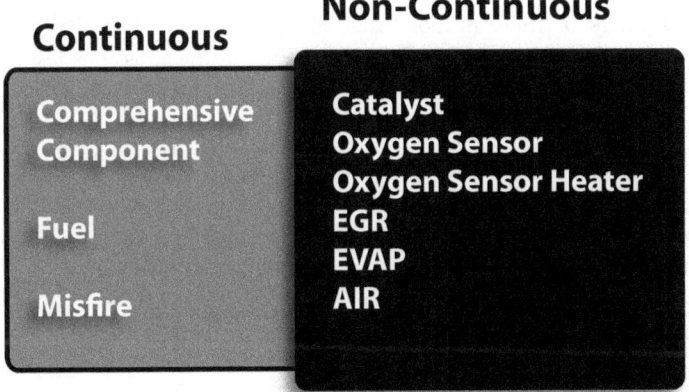

**Continuous**

- Comprehensive Component
- Fuel
- Misfire

**Non-Continuous**

- Catalyst
- Oxygen Sensor
- Oxygen Sensor Heater
- EGR
- EVAP
- AIR

# Monitors

### Non-Continuous Monitors

Non-continuous monitors are different from continuous monitors. If conditions are right and they complete an entire diagnostic test during a "trip," they will **stop** running until the next trip. (A **trip** is a set of carefully defined operating conditions that must be present for a monitor to run. See page 100 for more on trips.)

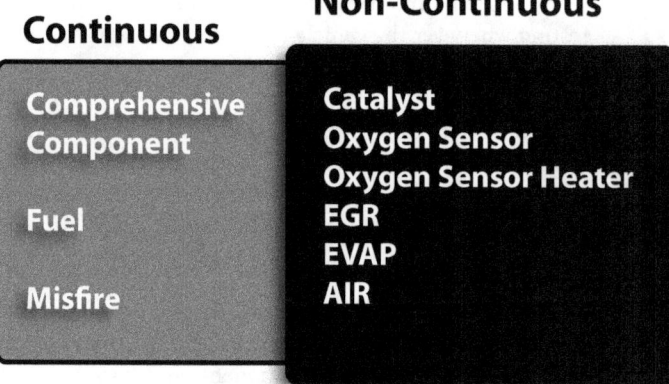

**Continuous**

Comprehensive
Component

Fuel

Misfire

**Non-Continuous**

Catalyst
Oxygen Sensor
Oxygen Sensor Heater
EGR
EVAP
AIR

Here are the continuous and non-continuous monitors. Remember, monitors test the vehicle during normal operation. No special procedure is needed to make them run.

Important! After non-continuous monitors run to completion one time, they are displayed as "ready." Even more importantly, they will stay "ready" unless emissions data are erased with a scan tool command or by a loss of battery power at the PCM.

You don't need to keep running monitors over and over again to *keep* them listed as "ready" on the scan tool. (That said, there are some goofball cars out there that make a liar out of the last statement, but the emissions test station has a list of these troublemakers and deals with them on an individual basis.)

> If monitors are reset to incomplete, the only way to change them back to complete is by driving the vehicle in such a way that the monitors can run again.

The Auto Emissions Bible

## Scheduling OBD II Tasks

Forgive the cartoon, but this little guy has been with me for many years, and he's a bit of a favorite. He represents the scheduling and record keeping department in the OBD II software suite. Some call him the **Diagnostic Executive**, others call him the **Task Manager**. The latter has always seemed to be more roll-up-your-sleeves relevant, so that's what we'll call him here.

Our Task Manager likes things done in a pre-defined order. He schedules tests based on operating conditions; runs monitors so they won't conflict with one another; stores all test data for future reference; stores and erases DTCs when necessary; controls the MIL; and records all information about monitor test scores and their completion status.

He may not look like an action hero, but don't underestimate his powers. He's the guy who decides if you get new license plates in areas that perform scan tool emission tests!

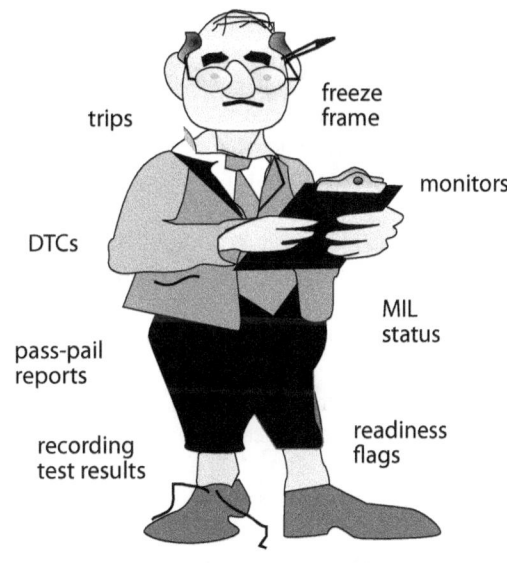

trips

freeze frame

monitors

DTCs

MIL status

pass-pail reports

readiness flags

recording test results

**Task Manager**

- We can't alter his routine, but we can learn how to use the data he stores.

- We cannot make him pass a monitor he thinks should fail, but we can learn how to look at the data he uses to make those decisions.

Throughout the rest of the book, as we look at monitors and discuss what it takes to get a vehicle to pass an emissions test, keep the Task Manager in mind. He is the symbol of the ordered test process programmed into OBD II vehicles.

# Monitors

## Running Monitors

So you want to take your car to the test center, but a dead battery has erased all emissions data and reset all monitors to incomplete. To get the Task Manager to change them back to "ready" so you can get your plates renewed, you'll need to drive the vehicle and complete *trips* for each of the monitors.

**A *trip* is a set of operating conditions that allow a monitor to run.** Each monitor has its own trip definition. Monitors can be finicky rascals; they run only when conditions are right, based on rules of engagement built into the car computer.

### Generic Drive Cycle

Here's a generic drive cycle that runs many monitors. If this doesn't complete enough of the monitors to get the vehicle through the emissions test, you may need to look up the exact drive cycle in a manufacturer repair manual or in a drive cycle reference manual, and then follow it to the letter.

**Step 1.** The vehicle must be sitting for 6-8 hours before the test, without a start. (This is primarily for the EVAP monitor in some makes that require a 6+ hour cold soak as part of the monitor enabling criteria.)

**Step 2.** Connect a scan tool. Start and warm the engine to normal operating temperature before driving it.

**Step 3.** Drive the vehicle for 10 minutes at highway speeds.

**Step 4.** Drive the vehicle for 20 minutes in stop and go traffic with at least four idle periods.

- Do **not** turn the ignition off at any time during the cycle.

- Take a passenger along to watch the monitor status display as you drive. Your passenger can watch the monitor readiness screen and inform you when monitors have run to completion.

- **Caution:** Some monitors will not update the scan tool display until after the ignition is turned off. And some monitors run only after the engine is shut off. Example: Some Chrysler oxygen sensor heater monitors run only after the ignition is switched off.

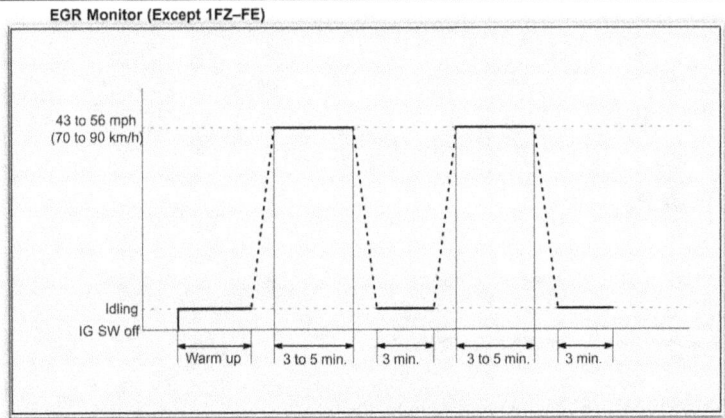

EGR Monitor (Except 1FZ–FE)

43 to 56 mph (70 to 90 km/h)

Idling
IG SW off

Warm up | 3 to 5 min. | 3 min. | 3 to 5 min. | 3 min.

This is a drive cycle diagram for a Toyota EGR monitor. Note how similar it is to the generic drive cycle: Warm up, acceleration, cruise, and deceleration are common parts of many drive cycles. Tests that run after engine shut down have very different enabling criteria.

## It's Fixed Now. Right?

Before scan tool emissions tests, many "fixed" vehicles were suitably lobotomized with a scan tool to erase DTCs and turn off the MIL, and then sent down the road. As monitors ran to completion, however, onboard monitors returned a verdict of their own about the effectiveness of a repair.

This led to an annoying tendency for "fixed" cars to return the next week with the MIL on for an entirely new DTC. Lesson learned? **The PCM has the final say about whether the MIL comes on, or stays off.**

Why does this happen? Some monitors are suspended for certain vehicle conditions. And one of the conditions that can suspend monitors is a DTC previously stored in memory. Some monitors refuse to run at all until other DTCs go away. You repair one problem; another monitor runs and stores a brand new DTC.

How about an example? You've fixed an oxygen sensor problem, you erase codes and slam the hood. Little do you realize that the catalyst efficiency monitor has not been running for weeks due to the oxygen sensor DTC that was previously stored in memory. Now that the oxygen sensor is fixed and its DTC is gone, the catalyst monitor runs again. Ooops! It fails the catalyst for low efficiency, storing a new DTC.

Back to the drawing board.

## Monitor Review

- Monitors test the vehicle during normal operation.

- OBD II vehicles have multiple monitors.

- Monitors run when conditions are right for them to run.

- There are two kinds of monitors: continuous and non-continuous.

- A trip is a set of operating conditions that allow a monitor to run.

### Added Thoughts About Monitors

Whether you are an auto repair professional or skilled prosumer, you have some decisions to make about monitors.

**1) Pause before erasing DTCs.** If the MIL was on because someone left the gas cap dangling from its tether after a fill up, reinstall the cap properly, and drive the car. Odds are, the PCM will test for leaks again and turn off the MIL. Depending on make, model, and equipment, this can happen in a day or two.

**2) Think before erasing DTCs.** Yeah. We already said this. But there is another reason not to go wiping out those DTCs. Say your car failed the emissions test for a bad TP sensor (throttle position sensor). In some areas, a retest requires that the catalyst monitor be complete. But if you erase DTCs and that catalytic converter is borderline, the monitor may take days—or longer—for the monitor to run to completion. If your plates expire tomorrow, it is easier and much faster to drive the car and let the PCM turn off the MIL after three trips. That way the catalyst monitor will still be shown as *ready*, since it was never reset.

**3) Make sure the car passes the first time.** If you just replaced the battery in your car, make sure you have enough complete monitors to pass the test the first time. Retest standards may be tougher.

**Need More Help? See Chapter 11- Passing the Emissions Test.**

# EMISSION CONTROLS

# 7

# Emission Controls

## Powertrain Management

This section deals with devices designed to clean harmful emissions from good engines working at peak efficiency.

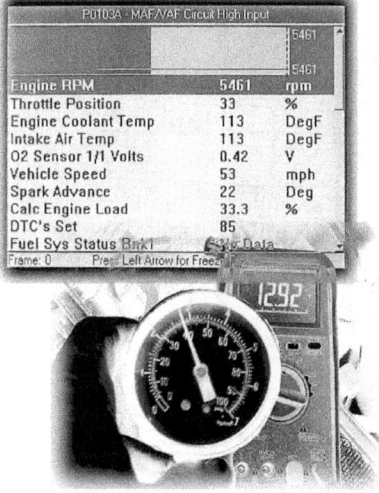

Begin with a thorough underhood inspection. Then use a scan tool, vacuum gauge, fuel pressure gauge, and digital multimeter to eliminate basic problems.

Emissions devices and mandatory tests sometimes leave us thinking that vehicle emissions are the main reason for improved engine designs and computerized closed loop engine management systems. That just isn't the case. The primary task of power-train management systems is to **improve performance**. More power. Better fuel economy. Less repair and maintenance.

An added benefit is that more efficient engines are cleaner to begin with.

The first line of defense in reducing emissions is to ensure that the engine is mechanically sound, and that the OEM engine management system is working as originally designed.

No engine is perfect. Some harmful emissions are produced by combustion in all engines. But don't concentrate on add-on emission control devices until you know for sure that the engine mechanical and engine management systems are in tip-top shape.

If you cringe each time someone says "check the basics," get over it. Check the basics. Carefully. That's where you will find most of the easy fixes.

See the test list on page 10.

## Closed Loop

Closed loop feedback systems measure exhaust oxygen; computers use this information to control fuel delivery. A regulated air/fuel ratio improves combustion efficiency and reduces harmful emissions.

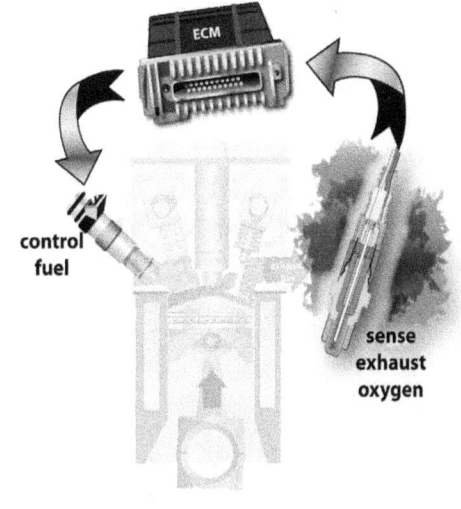

control fuel

sense exhaust oxygen

Traditional zirconia oxygen sensors cannot identify the exact air/fuel ratio. Instead, they measure two exhaust states: richer or leaner than stoichiometry.

- **High** $O_2$ sensor voltage indicates a **rich** mixture with little exhaust oxygen.

- **Low** $O_2$ voltage indicates a **lean** mixture (high exhaust oxygen).

This approach is far from perfect, producing inefficient fuel system overcorrections. Additionally, the traditional oxygen sensor switches only when the air/fuel ratio is in a fairly narrow band on either side of stoichiometry (a gasoline air/fuel ratio of 14.7:1). Excessively rich or lean mixtures are not accurately measured.

This is the reason traditional $O_2$ sensor feedback loops overshoot their targets. If you know this, use it: Fluctuating $O_2$ sensor voltage that repeatedly oscillates between approximately 200 and 800 mV at 2000 rpm is the single best predictor of overall fuel mixture control.

OBD vehicles use heated oxygen sensors that come to operating temperature in a minute or less. Look for new vehicles to enter closed loop as soon as the $O_2$ sensor comes on line, regardless of coolant temperature.

These DTCs are commonly associated with oxygen sensor and closed loop failures:

**P0130-P0167**

Slow $O_2$ sensor response or inactivity and oxygen sensor heater faults.

# Emission Controls

## Oxygen Sensors

The oxygen sensor does what its name suggests — it senses the amount of oxygen in exhaust gas. The amount of oxygen in the exhaust is normally an indication of the air/fuel ratio in the cylinder during the power stroke. Richer mixtures tend to consume more oxygen during combustion: lean ones, less.

Hello, computer?

We need more fuel...
no make that less fuel...
no, make that more fuel...
less fuel...
more fuel...
less..
more...

oxygen sensor

A good oxygen sensor is never happy. If exhaust oxygen content is high, it asks for more fuel. As soon as fuel is added and the mixture swings rich, it asks for less fuel. This process goes on constantly in closed loop, up and down, up and down.

Most oxygen sensors are **zirconium dioxide** (or zirconia, in common usage) voltage generators. They compare exhaust oxygen to atmospheric oxygen; if there's a big difference between the two, the sensor generates a higher voltage. Zirconia sensors generate voltage in a range of approximately zero to one volt, with higher voltage indicating a richer mixture.

Oxygen sensors are also referred to as *lambda sensors*. Lambda refers to a target air/fuel ratio of 14.7:1, a mixture that reduces most exhaust emissions without severely compromising engine performance. Oxygen sensor feedback is used by the PCM to maintain the air/fuel ratio near the 14.7:1 target, referred to as ***stoichiometry***. Since the oxygen sensor affects the fuel mixture (and the fuel mixture affects oxygen sensor voltage) we refer to this mode of fuel control as ***closed loop***.

Traditionally, oxygen sensor closed loop signal voltages dither back and forth from high to low—and back again—over and over. With the system in closed loop, this dithering establishes a range of alternating rich-to-lean transitions, bracketing the average voltage that represents the desired air/fuel ratio.

## Oxygen Sensors

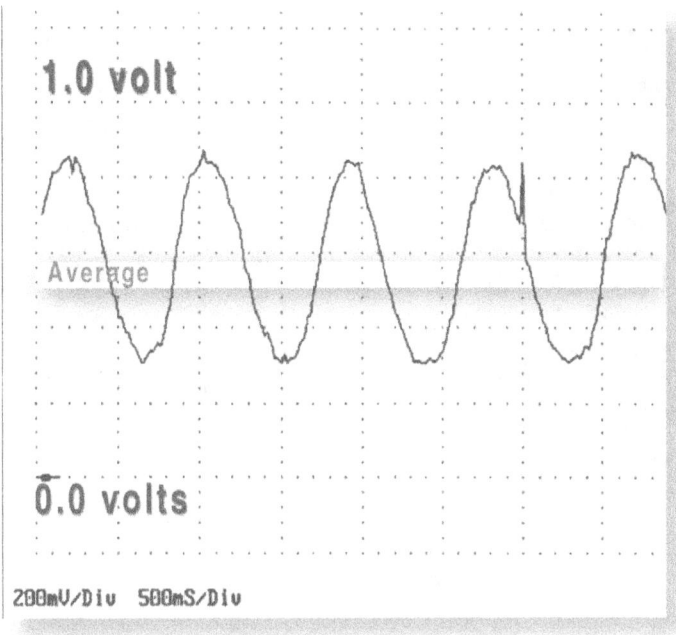

Use a scope connection with at least a 10 meg ohm impedance or the scope will load the circuit and skew the readings.

This is an oscilloscope display of an oxygen sensor signal in an engine running
at 1500 rpm. This sensor's signal voltage fluctuates quickly in a normal range,
with one or more complete high-low voltage cycles per second.

The minimum and maximum voltages suggest good sensor response. Note
that the average voltage is roughly centered between zero and one volt, indicating balanced fuel control.

An oscilloscope is the best way to monitor zirconia oxygen sensor signals. Scan
data and digital multimeters are too slow and do not add fine detail present
in a waveform. This will change when we get to wide range air/fuel sensors,
however.

Note the voltage and time settings for reference as a starting point for setting
up your scope.

# Emission Controls

## The AF Sensor or Wide Range Air/Fuel Sensor (WRAF)

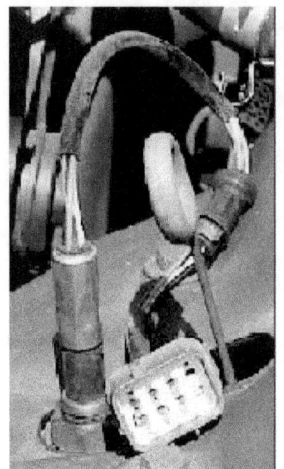

A new generation of oxygen sensors is at work in vehicles from several manufacturers. These are referred to as Air Fuel Ratio (AFR) sensors, AF sensors, LAF, and Wide Range Air Fuel sensors (WRAF). (More term warfare.)

Similar in appearance to a traditional zirconia sensor, the AF sensor is also a voltage generator. But that's where the similarity ends. We need to throw away the book and start over with this one.

Unlike traditional zirconia sensors, the AF sensor does not dither. It measures the *exact* air/fuel ratio, allowing more accurate fuel control.

It also has a far greater measurement range (hence the name *Wide Range*), identifying air/fuel ratios from 10:1 (rich) all the way to lean mixtures consisting of pure air!

---

• *Some* generic interfaces do not display AF sensor voltage correctly. This goes back to the original OBD II standards requiring $O_2$ sensor PID voltage to be displayed in a range between zero and 1 volt. (It's pretty tough to accurately display AF sensor voltage levels at 3.3 volts using a 0-1 volt scale.) Instead, what you'll see in **some** interfaces is a *percentage* of true voltage. To display the actual PCM PID voltage, in these cases, you'll need a scan tool with enhanced/factory software.

• Ready for more confusion? GM refers to the AF sensor as a Wide Range Air Fuel (WRAF) sensor and, in a GM vehicle, the WRAF PID is expressed as **lambda value**. The symbol for lambda is the Greek letter λ. (See page 139.)

GM suggests that lambda values on the Tech 2 may range from very rich, λ=0.75 (air/fuel ratio 11.05:1 ) to a "lean enough to breathe," λ=3.999 (air/fuel ratio 58.8:1).

• AF sensors operate at 1200°F, compared to 600-700°F for traditional thimble style zirconia sensors. AF sensors look a lot like traditional sensors, and may have 4, 5, 6, or even 7 wire harnesses.

---

# Emission Controls

## AF Sensor (WRAF) Characteristics

Here are a few characteristics of the AF sensor that make it unique, compared to traditional oxygen sensors:

**Wide Range Air/ Fuel Sensor**

- **The sensor is a voltage generator capable of reversing its current polarity.** When the fuel mixture is rich (low exhaust oxygen content), sensor current polarity is negative. Lean mixtures create a positive polarity current. This is an extremely low milliamperage. Direct current measurement is usually impractical.

- **Use your scan tool to monitor the AF sensor.** Unlike zirconia sensors, PID or Mode 6 data are easier to get and more useful that direct measurement with scope or meter. Note: AF sensor PID voltage doesn't come directly from the sensor; it is generated by the PCM in response to changes in sensor current.

- **Expect to see vehicles use both types**, with an upstream AF sensor for primary fuel control and a downstream zirconia sensor for catalyst efficiency and sometimes, to fine tune fuel control.

| ✓ Bank 1 - Sensor 1 (Wide Range O2S) (mA) | WRERB1S1 | 1.0085 | Ratio |
|---|---|---|---|
| ✓ Bank 1 - Sensor 1 (Wide Range O2S) (mA) | WRO2B1S1 | -0.0312 | mA |

| ✓ O2 Bank 1 - Sensor 2 | O2B1S2 | 0.7100 | V |
|---|---|---|---|
| ✓ O2 Bank 1 - Sensor 2 | FTB1S2 | 5.4687 | % |

This Honda Fit datastream capture shows both AF sensor and downstream sensor PIDs. The AF sensor value is given in mA (milliamps). **Ratio** indicates lambda. This rich mixture is creating a negative current flow from the AF sensor. The downstream sensor is a traditional zirconia sensor; its output is displayed as a voltage between zero and 1 volt.

| OBD Monitor ID (OBDMID) | Test ID (TID) | Test Value | Min Limit | Max Limit | Units |
|---|---|---|---|---|---|
| $01: A/F Sensor Monitor (Bank 1) | $80: B1S1 Activation Time | 9.800 | 0.000 | 40.000 | sec |
| $01: A/F Sensor Monitor (Bank 1) | $87: B1S1 Element Resistance | 39.000 | 0.000 | 250.000 | Ohm |
| $01: A/F Sensor Monitor (Bank 1) | $83: B1S1 Sensor Current Value | -0.016 | -1.230 | 127.996 | mA |
| $01: A/F Sensor Monitor (Bank 1) | $84: B1S1 Decel. Input Value | -1.625 | -2.172 | -0.500 | mA |
| $01: A/F Sensor Monitor (Bank 1) | $86: B1S1 Sensor Output Value | -0.129 | -2.254 | 127.996 | mA |

Same vehicle. Here we see the Mode 6 test results for the same AF sensor. This vehicle gives us two oxygen sensor diagnostic tools: datastream and Mode 6, both available through the generic interface. (See **Chapter 12- Mode 6** for more.)

# Emission Controls

## Catalytic Converter Chemistry

Catalysts promote a chemical reaction that converts harmful engine emissions into harmless water, carbon dioxide, and nitrogen. They do this by *reduction* and *oxidation*.

### Reduction

What's going on inside the catalytic converter? Most modern vehicles use a **three-way** catalytic converter that cleans HC, CO, and NOx. There are separate sections for oxidation and reduction of exhaust gases. The first section is called the "reduction bed." Reduction is the opposite of oxidation: it **removes** oxygen atoms. In this case it **removes** oxygen from NOx, allowing the nitrogen to reform as free nitrogen ($N_2$).

Reduction removes the oxygen from NOx and adds it to CO to produce non-toxic nitrogen and carbon dioxide.

$$NOx + CO \rightarrow N_2 + CO_2$$

### Oxidation

The second and third areas of the converter are the oxidation beds. Oxidation **adds** oxygen to compounds.

In oxidation beds, HC combines with $O_2$ to produce $CO_2$ and water ($H_2O$), while CO combines with $O_2$ to yield $CO_2$.

$$HC + CO + O_2 \rightarrow CO_2 + H_2O$$

# Emission Controls

## Catalyst Killers

Ideally, a catalyst should last as long as the vehicle. If you're forced to replace a catalyst, ask **why** it failed before bolting on a new one. Catalysts that die young do so from contamination and/or damage caused by overheating.

Common sources of contamination include engine oil, antifreeze, and silicone sealers that are not rated catalyst safe. Contaminants coat the catalyst's active materials, reducing chemical action and catalyst efficiency.

Catalytic converters will work as hard as they can. In fact, they will work themselves *to death*, if you let them. Overly rich conditions and engine misfires pump more CO and HC into the cat than it can handle. This increases chemical activity and heat to a point where the catalyst suffers a meltdown.

Never replace a catalyst without locating and correcting the conditions that destroyed it in the first place.

### Why can't the converter just clean up ALL the exhaust gases?

Catalytic converters work when the basic mixture is close to stoichiometry. Converter efficiency is best when lambda falls in a range of 0.97 to 1.03. The catalyst does a much poorer job of cleaning the exhaust when lambda falls outside that range, indicating a mixture that is far too rich or far too lean.

A word about replacement converters: Aftermarket converters are not required to meet the same specs as original equipment. If you feel you must use an aftermarket converter, you will usually get better and longer lasting results using the heaviest-duty universal converter listed by the supplier for the vehicle.

OBD II vehicles need OBD II-certified converters. For best results, use a CARB-certified replacement converter and, if maintaining long term OEM performance is important to you and your customer, select an OEM replacement.

# Emission Controls

## Secondary AIR Injection

Atmospheric air is injected under pressure into the exhaust or catalyst by an air pump, in some makes. This occurs primarily during engine warm-up following a cold start, to oxidize fuel left over from cold start enrichment, before it reaches the catalyst. The pump may also be activated at other times as part of an intrusive test to monitor other components.

For our purposes, it is important to remember to disable air injection when we evaluate exhaust gas $O_2$, since air injection can elevate $O_2$ levels.

**Caution:** Disabling air injection does not mean clamping off air injection hoses and tubes. This can permanently damage fragile plastic hoses and valves.

Instead, before taking tailpipe exhaust samples, use OEM recommendations to disable the air pump electrically, or to divert air injection to atmosphere through the pump air vent.

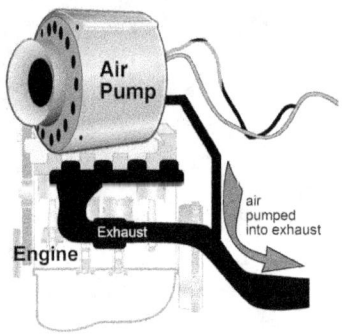

Belt-driven air pumps have largely been replaced by PCM-controlled electrical pump motors.

**OBD DTCs**

DTCs commonly associated with secondary AIR failures:

**P0410-419**
Miscellaneous AIR failures

## Positive Crankcase Ventilation

Positive Crankcase Ventilation is the granddaddy of add-on emission control devices. It may also be the most misunderstood.

The PCV valve recirculates cylinder blow-by gases to the intake manifold for burning. (Blow-by gases are those that force their way past the piston rings into the crankcase.) The PCV system replaces those blow-by gases in the crankcase with fresh air. PCV failures result in inadequate or excessive PCV flow rates, often caused by sticking or improperly calibrated PCV valves.

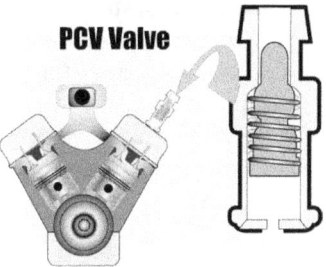

**PCV Valve**

- **Excessive PCV flow** may cause increased exhaust HC and CO if too much fuel vapor is drawn from the crankcase. PCV vacuum leaks may cause higher than normal exhaust $O_2$.

- **Insufficient PCV flow** allows blow-by gas to build up inside the crankcase. Excessive crankcase pressure may force oil past seals and gaskets, causing engine oil leaks.

Many vehicle emissions failures are the direct result of low quality, improperly calibrated replacement PCV valves. Be choosy when ordering new PCV valves. Use only high quality valves from a known, reputable source. When in doubt, use an OEM replacement. Installing the wrong valve may cause a vehicle to fail a tailpipe test that it would otherwise pass!

New PCV systems must have hose connections with strong mechanical attachments to keep them from being knocked off during normal maintenance. Some vehicles use very large PCV hoses that create a vacuum leak big enough to stall the engine if the hose is not properly connected.

Having a hard time finding the cause of an HC failure? Yank the PCV valve with the engine running: Leave the valve connected to the suction hose at the intake manifold, but pull the valve inlet away from the engine and let it draw fresh air. If HC drops significantly, the crankcase is probably saturated with raw fuel.

**OBD DTCs**

DTCs associated with PCV failures:

**P0171; P0174**
Fuel System Lean

**P0172; P0175**
Fuel System Rich

**P0505**
Idle Air Control System

# Emission Controls

## Evaporative Emissions

The emissions subsystem that has undergone the biggest change is the **Evaporative Emissions System**. Many of you know this as the EVAP system.

For many years, fuel vapors from the gas tank have been collected and stored inside a canister filled with activated charcoal. At the correct time, a command from the PCM opens a purge valve connected to engine vacuum. This draws vapors from the canister into the engine, where it is burned.

### Early EVAP Systems

Charcoal canisters found in early EVAP systems have a vent to atmosphere that is always open. The vent allows fresh air to be drawn into the canister as vapors are purged. It also acts as a safety valve, relieving excess tank pressure generated by high ambient temperatures. (Fuel vapors are not vented, only air; gaseous HC is captured by the charcoal canister and held until purged to a running engine.)

Purge is enabled when the engine can easily burn the additional fuel vapors, commonly during steady cruising with the engine at normal operating temperature, although some vehicles will purge at idle. Excessive or incorrectly timed purge can cause hard starting, stalling, or general poor performance.

Common problems with these systems include:

- **A fuel soaked canister that contains liquid fuel instead of fuel vapors.** This results in an over-rich air/fuel ratio. EVAP systems have a liquid valve/trap designed to prevent liquid fuel from entering the canister through the vapor line between the tank and canister. A leaking valve or overfilling the fuel tank, however, can saturate the canister with liquid fuel. This results in an extremely rich fuel mixture when the purge valve opens.

- **A sticking or inoperative purge valve.** Opening the purge valve at the wrong time causes driveability problems. A purge valve stuck in the open position can drive mixtures rich or lean, depending on how much (or how little) fuel vapor is stored in the canister.

- **Clogged canister/vent filters.** If the canister can't breathe properly, it may be difficult to fill the gas tank. A clogged vent can also result in a high enough vacuum inside the fuel vapor system to collapse the tank walls as the tank empties.

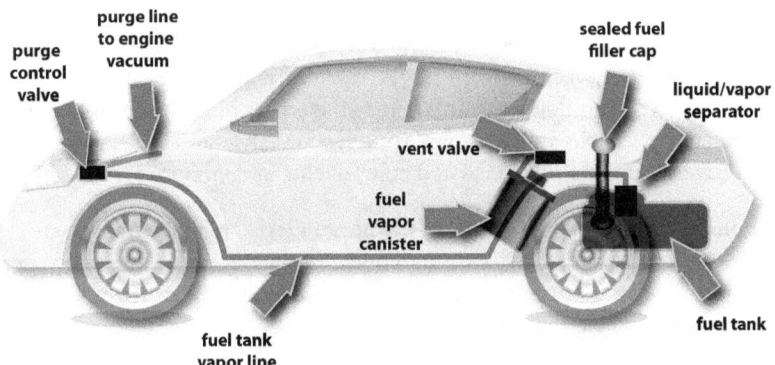

This illustration shows common EVAP system components and their general locations. You may need a component locator and a little digging to unearth and test certain sensors and actuators. Bidirectional scan tool commands can save time by energizing actuators to test their function, eliminating a lot of needless vehicle disassembly.

The PCM tests the Purge Valve electrical circuit for shorts and opens.

It may also monitor purge flow using several methods that commonly include some combination of: direct measurement with a purge flow sensor; by watching IAC (Idle Air Control) counts; by watching the oxygen sensor to determine if the air/fuel ratio changes when the purge valve opens; by watching signal changes from the fuel tank pressure (FTP) sensor.

## OBD DTCs

DTCs commonly associated with EVAP purge failures:

**P0171; P0174** - Fuel System Lean

**P0172; P0175** - Fuel System Rich

**P0440** - Fuel System Fault

**P0441** - No EVAP purge

**P0442** - EVAP system leak

**P0443** - Incorrect purge flow/ purge control valve fault

**P045x** - EVAP system leaks

# Emission Controls

## Leak Detection Simplified

Current EVAP systems check for fuel vapor leaks in the entire fuel containment system; tank, lines, hoses, fill cap, etc. A common leak detection strategy is to apply vacuum to the EVAP system, and then seal the system to see if the vacuum holds. As different as the terminology and component placements in various vehicles may seem, vacuum style leak detection systems have several main features and components in common:

• **A purge control valve, similar to the purge valves we've seen for years.** These valves are normally *closed*, and open during purge to draw fuel vapor from the canister or create a test vacuum inside the system. The purge valve closes during leak tests.

• **A vent solenoid, added to the canister or canister vent line.** Unlike the purge valve, the canister vent is normally *open*. The PCM closes the vent *only* when it needs to seal the system for leak testing.

• A **fuel tank pressure sensor**. The Fuel Tank Pressure (FTP) Sensor, aka the Fuel Tank Pressure Transducer (FTPT), is calibrated to inches-of-water. It measures both pressure and vacuum.

## Leak Detection Strategy

Here are the generic essentials of vacuum leak detection. You will see variations on this theme:

• KOER (Key-On-Engine-Running), the PCM **closes** the canister vent and **opens** the purge valve.

• Engine vacuum creates **low pressure** (small vacuum) in the fuel vapor system.

• When the fuel tank pressure sensor indicates that a small vacuum is present inside the fuel system for a specified time, the system passes the large leak test. If no test vacuum can be drawn, a large leak trouble code is stored.

• The purge valve closes, trapping test vacuum in the system (the canister vent is already closed).

• The PCM watches the pressure sensor to see if the vacuum holds. If vacuum "decays" too rapidly, a small leak diagnostic trouble code is stored.

# Emission Controls

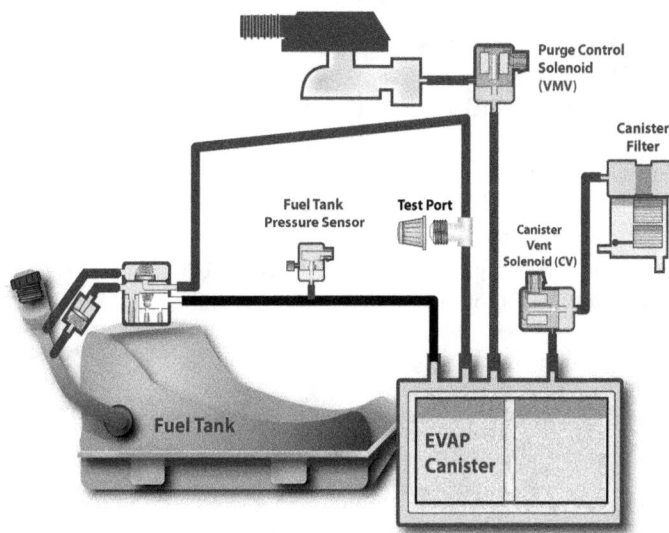

A system overview of an EVAP system showing major components, and their general locations. Vapor leaks may be caused by loose or damaged components or a loose gas cap.

Note: Some test programs still use a gas cap check. This may be as simple as a visual inspection to see if the cap is missing, or a pressure test to locate leaking caps.

DTCs commonly associated with EVAP system leaks and system components:

**P0440** - EVAP no/low flow during purge; no tank vacuum; purge fault

**P0441** - EVAP no purge flow

**P0442** - EVAP system small leak

**P0443** - Purge solenoid/circuit fault

**P0446; P0449** - Vent valve/solenoid circuit faults

**P0450-P0454** - Fuel Tank Pressure Sensor fault

**P0455** - EVAP system large leak

# Emission Controls

## Leak Detection Pumps

Some OEMs, including Chrysler and VW, have used *pressure* instead of vacuum decay to test fuel vapor systems for leaks.

A small vacuum-operated Leak Detection Pump (LDP) produces a test **pressure** inside the fuel tank, canister, and all fuel vapor lines. Then it stops. If system pressure drops too quickly, the pump runs again. The PCM runs a stop watch on the pump to measure the intervals between pump run cycles. If the pump cycles too often, a leak is indicated and a DTC is stored.

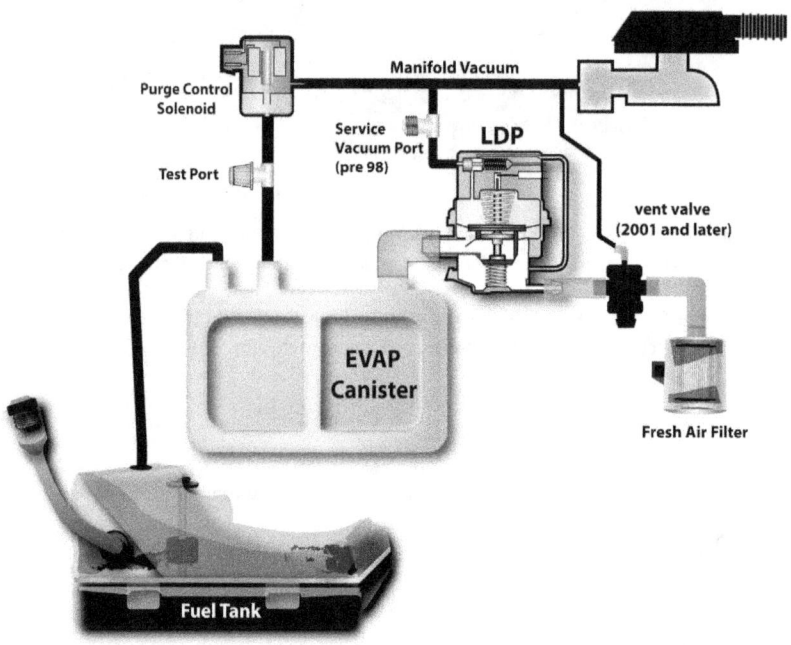

The pressure generated inside the system is very small; a fraction of one psi. The test pressure standards are so small that *pounds-per-square-inch* is too large a measurement scale. Instead, all system pressures are measured in *inches-of-water*.

The LDP applies a test pressure to the system of about 7 inches-of-water (inH20), a pressure equal to about ¼ psi.

## Natural Vacuum Leak Detection

Carmakers currently favor **Natural Vacuum Leak Detection**. It is simpler, cheaper, and has fewer components. Its big benefit is that it runs the leak test **after** the car is parked. This eliminates a common problem in a moving vehicle: fuel slosh. Gasoline sloshing around inside the tank of a moving vehicle generates more vapor pressure that can fool the EVAP monitor. Running the monitor under stable conditions improves test accuracy.

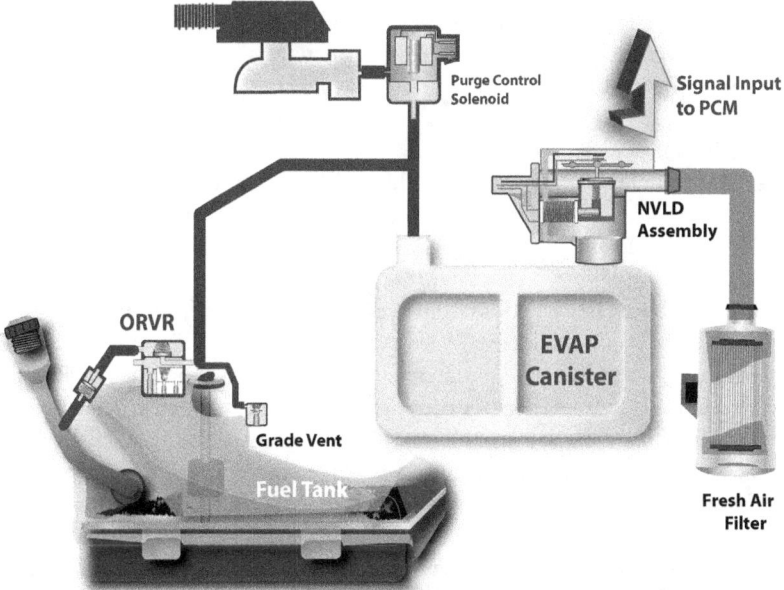

NVLD works by monitoring changes in vapor pressure caused by changes in vapor temperature. Unlike EVAP test systems before it, NVLD works after the vehicle is shut down. Here's how:

• The canister vent closes to seal the system. Pressure is monitored.
• If no leaks are detected the system waits for the cooling fuel to contract and create a fuel vapor system vacuum.
• A pressure switch closes when 1 inH$_2$O vacuum is reached, indicating that there is no leak. The system stores a "pass" when the switch closes.
• If the switch does not close, the test is inconclusive; no pass is recorded.
• Failure to record a pass over a designated time period stores a DTC.

# Emission Controls

## Ford EONV

Ford calls its natural vacuum leak detection strategy the **Engine Off Natural Vacuum (EONV)** test. This monitor is controlled by a small energy-efficient microprocessor located inside the main PCM that runs *after* the vehicle is shut down. The first use of EONV is in 2005 F-series super duty 5.4 and 6.8L trucks.

EONV hardware is similar to components found in vehicles that use a conventional engine vacuum leak test method. The difference? Instead of using engine vacuum, the monitor measures the natural vacuum (or pressure) inside the fuel system following shut down to test for leaks.

Here's how it works:

**1) The vehicle shuts down.** After the ignition is switched off, the canister vent solenoid (CVS) remains open to allow tank and atmospheric pressures to equalize.

**2) The PCM watches EVAP system pressure.** If the Fuel Tank Pressure Transducer (FTPT) detects EVAP system pressure of 1.5 inHg (0.05 psi) or more, the test software assumes that fuel in the system is very volatile (generating a lot of fuel vapor pressure). Since excess pressure makes the test unreliable, the processor aborts the tests. (Remember, the canister vent is open. If the system can generate a positive pressure with the vent open, it's assumed that the fuel is volatile, creating a large amount of vapor.)

**3) The PCM closes the canister vent (CVS).** Since the VMV (Vapor Management, or Purge Valve) is already closed, the system should be sealed. The microprocessor continues to look at feedback from the fuel tank pressure sensor. If the EVAP system is leak-free, pressure should change as tank temperature changes.

**The PCM continues to watch EVAP system pressure.** If the system has a leak, pressure will change very little, or not at all. If the PCM sees a large change in pressure (either a positive pressure or vacuum), it assumes that the system is sealed, and the test passes. The amount of change required for a pass depends in part on ambient temperature and fuel level.

**4) If a positive pressure is recorded, but it is not high enough for a pass, the first test is repeated.** The canister vent opens again to vent excess pressure from the system.

## Ford EONV (cont)

5) **The PCM closes the canister vent again to reseal the system.** The system is monitored again for pressure or vacuum. If a large enough pressure change is measured, a passing grade is recorded for the test. The test times out in 45 minutes. If a pass is not recorded in this time, the test fails.

6) **The test is complete.** The PCM re-opens the canister vent and all electrical loads are turned off to prevent continued battery drain. At the next engine start, test values and test results are sent to the PCM.

To prevent false trouble codes, the results of four key-off test cycles are averaged to store the first fault. Since this is a two-trip code, about 8 key-off test cycle failures are needed to store a DTC and illuminate the MIL.

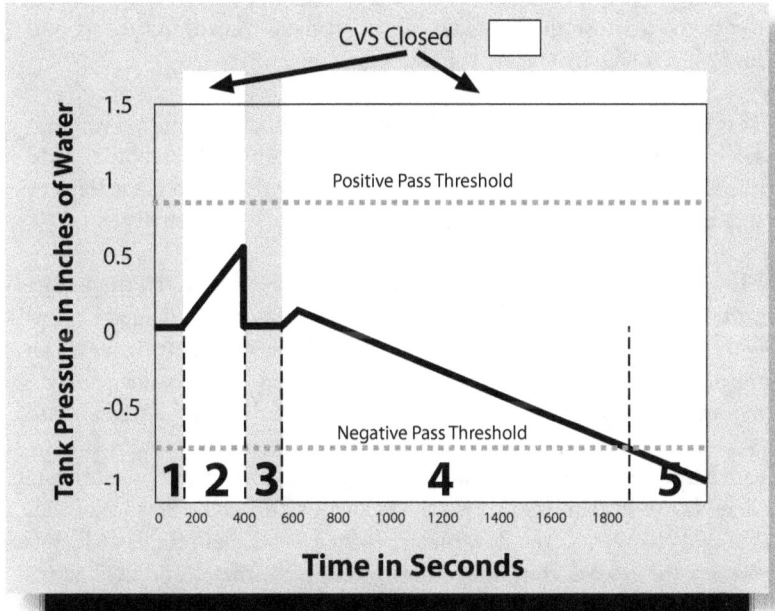

The steps in EONV testing are shown in this graph. Here we see a negative (vacuum) pass, although there is a positive pressure threshold, as well.

- Tank pressure above zero indicates pressure, in inches of water.
- Tank pressure below zero indicates vacuum, in inches of water.
- For reference, one psi equals 27.67 inH₂O (inches of water.)
- Test pressures used to locate EVAP system leaks are very small.

# Emission Controls

## EVAP Leak Testers

Once you know you have a leak, you need to locate and eliminate it. A number of dedicated EVAP leak detection machines are available. They locate leaks using (or combining) a flow meter, a pressurized inert test gas, and a smoke generator. (See pages 124-125 for more.)

### Machine Features

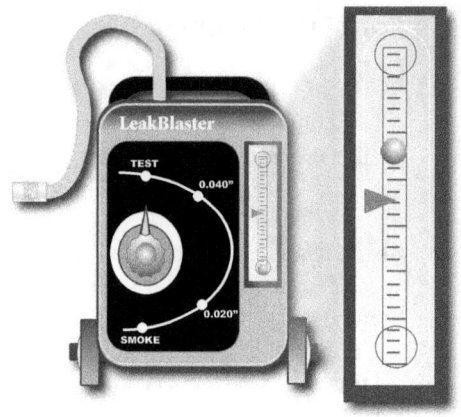

- **Adapters** - A good EVAP leak detection machine should include the necessary EVAP port adapters to connect the machine to vehicle EVAP test ports. Remove the Schrader valve (left hand thread) in the test port for best results when using a smoke machine.

- **Test Mode** - The machine should have a test mode that allows you to safely pressurize the EVAP system using nonflammable carbon dioxide or nitrogen gas. We recommend that you use only these two gases for pressure testing, and recommend $CO_2$ if you have a 4-5 gas analyzer.

- **Flow Meter** - The flow meter is connected in series between the test gas bottle and the vehicle EVAP system. If the ball in the flow meter floats with the vent valve closed, there's a leak in the system. Set the meter for the allowable leak (0.020 or 0.040 inch). Making the adjustment may require you to install a meter orifice or to turn a dial on the tester console.

Ready? Let's pressurize the system. Close the vent valve and pressurize the system. If the system is leak-free, the ball will settle to the bottom of the calibrated chamber in the flow meter. If there is a leak in the EVAP system, however, test gas will continue to flow through the meter, and the ball will float in the chamber. The greater the leak, the higher the ball floats.

**CAUTION:** Close the vent valve using a scan tool command, whenever possible. If you must energize the vent valve solenoid manually, make sure you identify hot and ground and *observe circuit polarity*. We don't want to damage the vent control circuit (including the PCM) by grounding the hot wire! When in doubt, pull the circuit wiring diagram and use it.

**Gas Tank Fill Neck Adapters -** On vehicles with no EVAP test port, gas tank fill neck adapters provide an easy access point for pressure and smoke tests. If you test through the fill neck, don't forget to test the cap separately!

**Low Pressure Gauge** - A low pressure gauge is useful for comparing actual system pressure to the tank pressure sensor PID.

## Special Notes:

- **Some leak detection systems come with UV dye and a black light.** This is especially appealing to those who like to make those leaks looks like a Peter Max poster, just like on CSI!

- **Since the canister vent is normally open, you'll need to close it before leak testing.** This can be done either through a scan tool command (on some makes, if you have the correct scan tool software), or manually by applying power and ground to the *correct* canister vent solenoid terminals. Disconnect the canister vent electrical connector first, and identify power and ground to avoid reverse-polarity voltage that may damage sensitive solid state components.

- **CAUTION: Never apply more than one psi of test pressure to any EVAP system.** To do so is dangerous and may cause system damage or personal injury. Additionally, in many cars, you will damage the FTP sensor irreparably.

- **CAUTION: Some corroded fuel filler necks have holes in them too small to see with the naked eye; they will leak fuel vapor pressure to atmosphere.** There's no guarantee you'll see fuel leaks during a fill up, especially if the holes are located at the top of the neck and encrusted with road salt and dirt. Spray them with soapy water after applying normal test pressure to the system.

# Emission Controls

## A Sample EVAP System Leak Test Procedure

Leak detection methods vary. Some of us locate leaks by pressurizing the system with nitrogen while checking hoses and fittings with an ultrasonic leak detector or soapy water. Some use $CO_2$ and a gas analyzer. These methods are not always successful, however, especially when leaks occur in hoses and fittings that are hidden in frame rails or above the fuel tank.

The "smoke machine" (a big favorite for finding leaks in manifold gaskets, heater control vacuum lines, exhaust systems, etc.) has been adapted for use in dedicated EVAP leak detection units.

Let's look at ways to use EVAP leak detection devices. Please look at the next page and follow along as we walk through a typical leak test procedure.

1) **Verify the DTC that has illuminated the MIL.** Look up the exact DTC definition provided by the vehicle manufacturer for the make, model, and year. Use that information to narrow the search area and to determine the maximum allowable leak (0.020 or 0.040 inch).

2) **Locate the EVAP test port.** (Note: Not all vehicles have these ports. In some cases, you need to use a gas tank fill neck adapter or install your own test tee.)

3) **Remove the Schrader valve from the test port.** This is a left hand thread, so screwing the valve *clockwise* removes it. Removing the valve allows a free flow of test gas or test smoke into the system. Leaving the valve in place breaks up test smoke, reducing its density and making it harder to see as it exits a leak.

4) **Follow the equipment manufacturer's procedure to adjust the test equipment for the size of leak to be detected (0.020 or 0.040 inch).** In some machines this requires the installation of the correctly sized orifice restriction. On machines with flow gauges, like the one shown here, adjust the reference needle on the flow meter to indicate the maximum allowable leak size.

5) **Close the canister vent.** If possible, do this with a command from your scan tool. The valve can also be closed by applying power and ground to the correct terminals of the Canister Vent electrical connection. Observe circuit polarity and make sure gas cap is screwed on tightly.

6) **Apply test pressure to the system.** (**CAUTION**: Never apply more than one psi of test pressure to any EVAP system. (It's dangerous and may cause system damage or personal injury.) Nitrogen or $CO_2$ are good choices since neither support combustion. $CO_2$ leaking from the system can also be tracked with your gas analyzer.

7) **Look at the flow gauge.** If there is a leak, the ball will float in the chamber. The higher it floats, the larger the leak. The height of the ball indicates the leak size. If the ball does not rise at all, there is no leak. (If this happens, there's a chance that the Schrader valve you removed may be the source of your trouble code.)

The Auto Emissions Bible

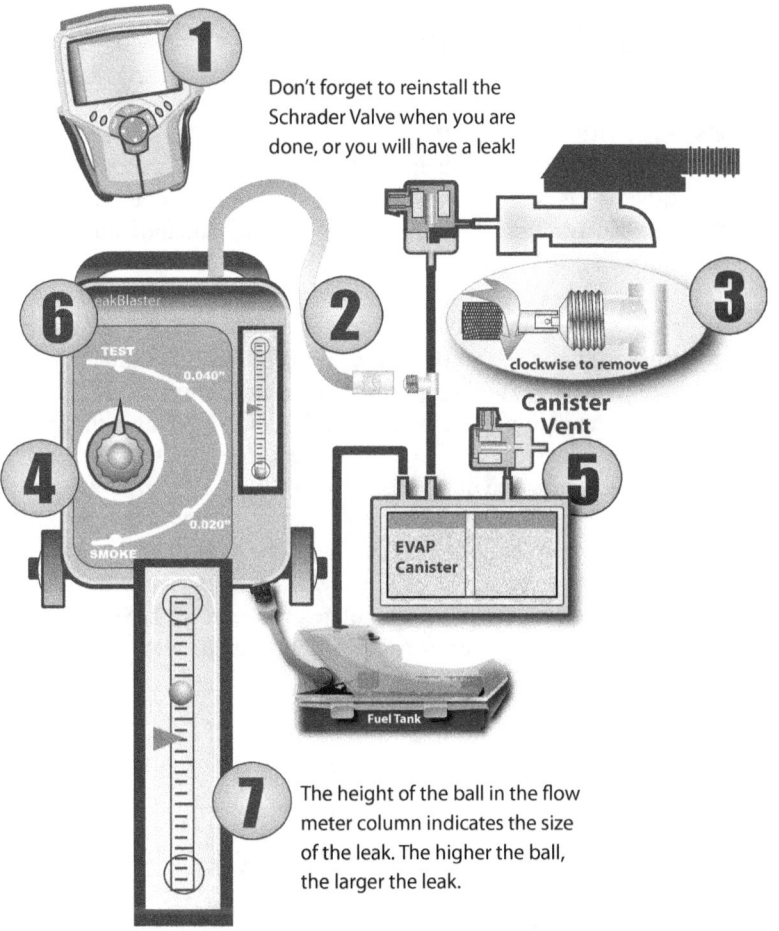

Don't forget to reinstall the Schrader Valve when you are done, or you will have a leak!

clockwise to remove

Canister Vent

EVAP Canister

Fuel Tank

The height of the ball in the flow meter column indicates the size of the leak. The higher the ball, the larger the leak.

• If the leak is not large enough to hear with the naked ear, introduce smoke into the system and look for telltale signs that will indicate the exact location of the leak. (Some leak detection systems allow you to add dye to the system that can be seen with an ultraviolet light.)

• The lower the fuel level in the tank, the longer it will take to fill the system. You may want to remove the gas cap until smoke comes out to determine when the system is filled with smoke.

• Use a strong spotlight to see small traces of smoke in dark places.

• Worth repeating: Some small inaccessible leaks can be located by pumping $CO_2$ into the system and probing around the EVAP system with your gas analyzer probe.

# Emission Controls

## Exhaust Gas Recirculation

EGR reduces NOx emissions. When the EGR valve opens, a metered amount of exhaust gas is routed *back* to the combustion chamber. Adding the inert exhaust to the intake air slows the rate of oxidation during combustion.

As campers know, throwing a small amount of ashes on top of a campfire makes it burn slower and cooler. By adding a small, controlled amount of already burned exhaust gas to the incoming air and fuel mixture, the EGR (Exhaust Gas Recirculation) system accomplishes much the same thing inside the combustion chamber. Cooler burning keeps peak temperatures below 2500°F where nitrogen burns, becoming NOx.

EGR is a difficult system to design and is one of the least robust emission control strategies in use. By nature, EGR valves and tubes clog with carbon residue from the exhaust gases they recirculate. Additionally, EGR must be timed and metered or engine performance suffers dramatically.

For more, see **Chapter 13: NOx.**

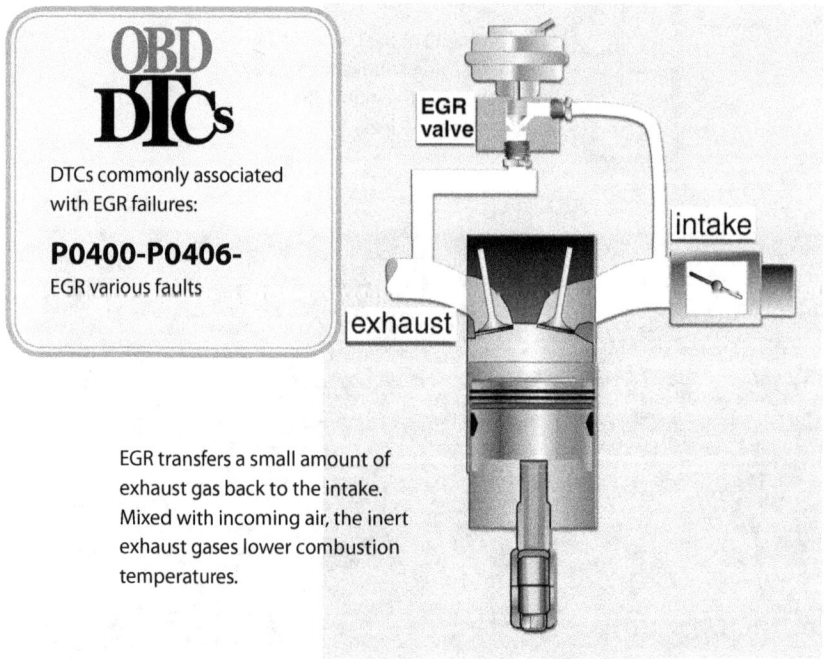

OBD
**DTCs**

DTCs commonly associated with EGR failures:

**P0400-P0406-**
EGR various faults

EGR valve

intake

exhaust

EGR transfers a small amount of exhaust gas back to the intake. Mixed with incoming air, the inert exhaust gases lower combustion temperatures.

## Exhaust Gas Recirculation

Some systems modulate EGR valve opening based on exhaust backpressure. This properly matches EGR flow volume to engine load.

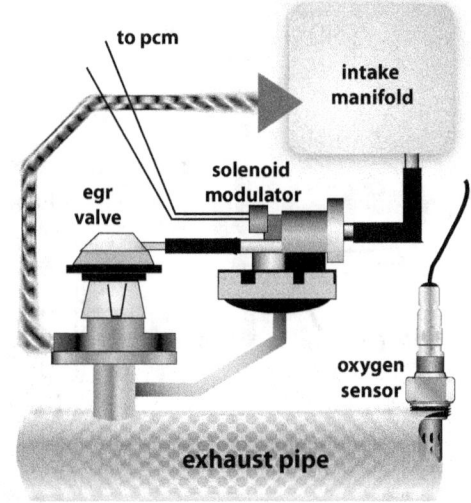

These EGR valves are vacuum-actuated through a PCM-controlled solenoid/modulator. EGR control vacuum vents to atmosphere until there is enough exhaust backpressure present at the base of the backpressure modulator. The combination of backpressure and control vacuum modulates EGR valve opening, based on engine load.

The PCM may test for exhaust flow by watching for shifts in exhaust oxygen as the EGR opens and closes, or by measuring EGR passage temperature.

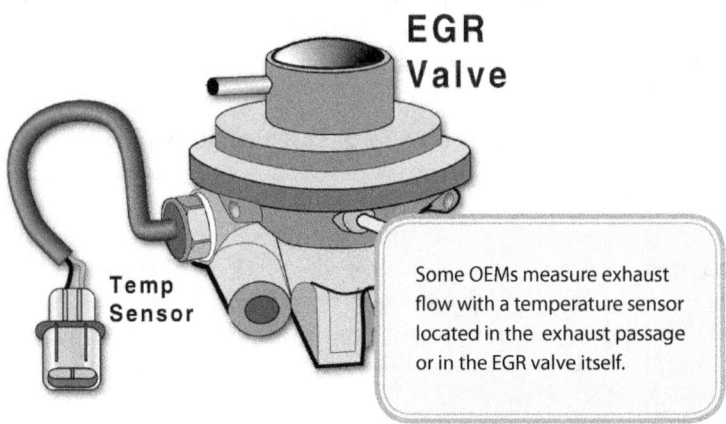

**EGR Valve**

Temp Sensor

Some OEMs measure exhaust flow with a temperature sensor located in the exhaust passage or in the EGR valve itself.

# Emission Controls

## DPFE (Differential Pressure EGR)

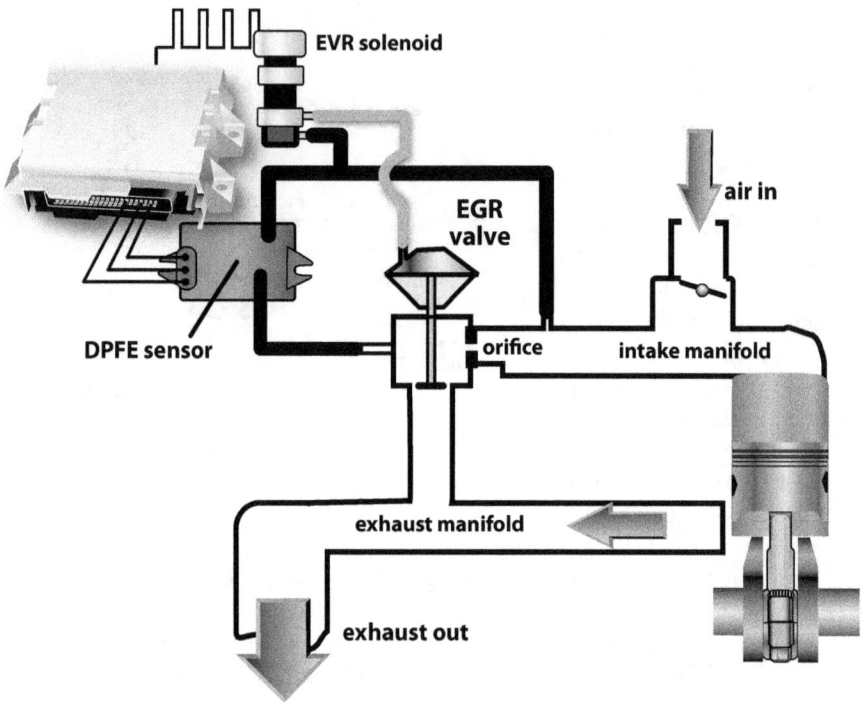

This image shows a variation on the original DPFE system. Note that one of the pressure sensing hoses is attached to the intake manifold. This allows the DPFE to provide manifold and BARO pressure inputs to the PCM, as well as EGR flow information. Might this be a good place to tee in your vacuum gauge? (See page 174.)

Many Ford vehicles have a pressure sensing **Differential Pressure EGR (DPFE)** that allows the PCM to regulate EGR flow rates, based on engine load.

A computer-controlled vacuum switching valve (VSV) applies control vacuum to regulate EGR valve opening. A pressure sensor monitors actual exhaust flow by measuring the pressure drop across a fixed orifice in a metering chamber located between the exhaust system and intake manifold.

The DPFE design and materials have been modified several times. It has a high failure rate, often due to carbon clogging or heat damage.

## Linear EGR Valves

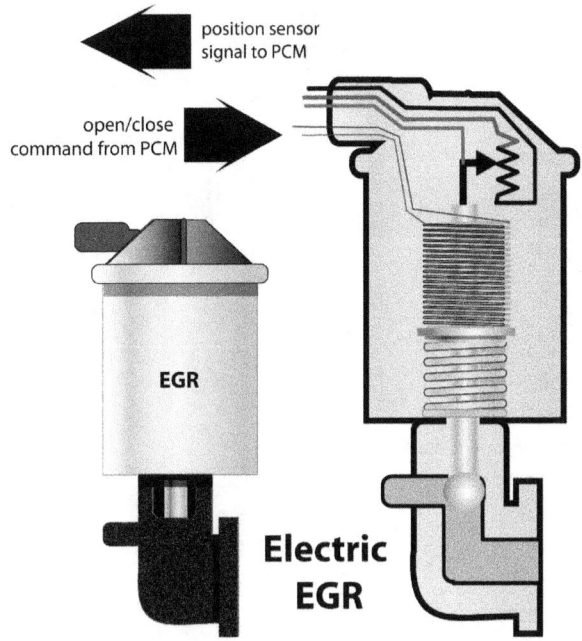

position sensor
signal to PCM

open/close
command from PCM

EGR

**Electric EGR**

EGR can be a real problem child for vehicle makers. By its very nature, the EGR system operates at very high temperatures, and since it recirculates exhaust and oil residue, it is susceptible to carbon clogging of both EGR valves and passages. Calibration errors are also common when valves stick wide open, fully closed, or out of position, allowing too much or too little exhaust gas to recirculate.

The electrically-operated **linear EGR valve** is common in many vehicles from various OEMs. A small stepper motor inside the EGR valve body opens and closes the valve on command from the PCM. A position sensor signals changes in valve position as it opens and closes. This lets the PCM compare the valve's *commanded* versus *actual* positions.

To verify flow, the computer commonly monitors MAP sensor voltage, looking for changes in intake manifold pressure that should accompany any increase in exhaust flow into the manifold.

These valves still stick, and many require periodic cleaning.

# Emission Controls

## VVT and EGR

Conventional EGR systems have often experienced rough running when the EGR opens at idle, or when a large load of exhaust gas is dumped into a single cylinder.

Rough running may also occur when exhaust gas fails to mix well with the incoming air/fuel charge. These problems commonly result from clogged EGR passageways, sticking EGR valves, or damaged vacuum lines.

In the mid-1990s, car makers introduced **variable valve timing** (**VVT**) as an alternative to conventional EGR.

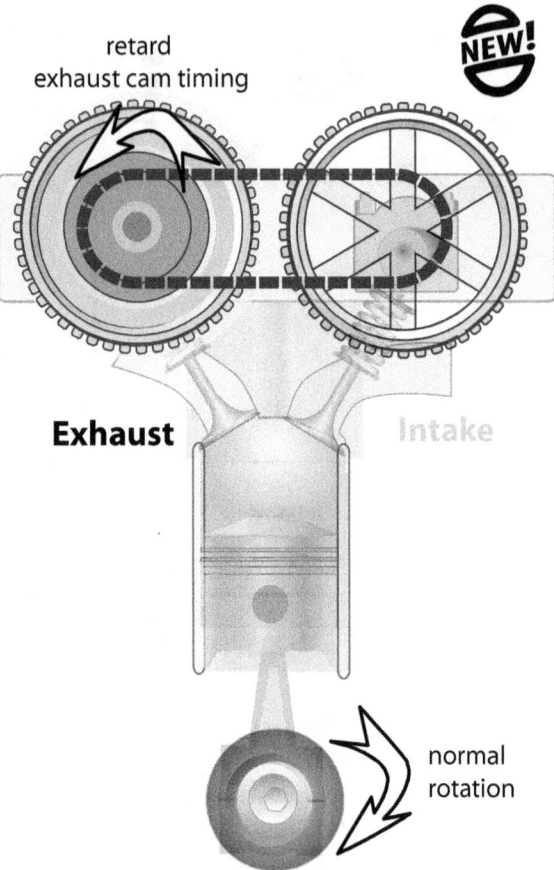

retard
exhaust cam timing

NEW!

**Exhaust**

Intake

normal
rotation

VVT/EGR retards cam timing, on the fly. Retarding the exhaust cam closes the exhaust valve *later*. This pulls a small amount of exhaust back into the cylinder during the intake stroke. The exhaust gas mixes better with the incoming rush of air and fuel. NOx is controlled, without the need for a separate EGR system.

VVT has another benefit. Leaving the exhaust valve open a little longer siphons fresh air into the cylinder faster. Its a phenomenon called "scavenging."

## VVT and EGR

**NEW!**

In VVT, a fluid-operated control valve rotates a cam timing mecha-nism built into the camshaft sprocket(s). Cam timing is adjusted in degrees to regulate the amount of exhaust gas that gets trapped inside the cylinder. Retarding the exhaust cam timing holds the exhaust valves open longer, to increase valve overlap. This mixes the incom-ing air/fuel charge and trapped exhaust more completely during the intake stroke. Trapped exhaust is precisely metered, and injector pulse-width and ignition timing are modified to reduce NOx, without affecting engine performance.

Common problems include sticking cam phasers, often due to oil sludge caused by lack of basic maintenance.

Other problems include the wrong oil viscosity, low oil pressure, and electrical or hydraulic phaser solenoid failures. Even the wrong oil filter can cause the system to malfunction.

Cases of improper reassembly during a timing belt job have also resulted in numerous driveability symp-toms.

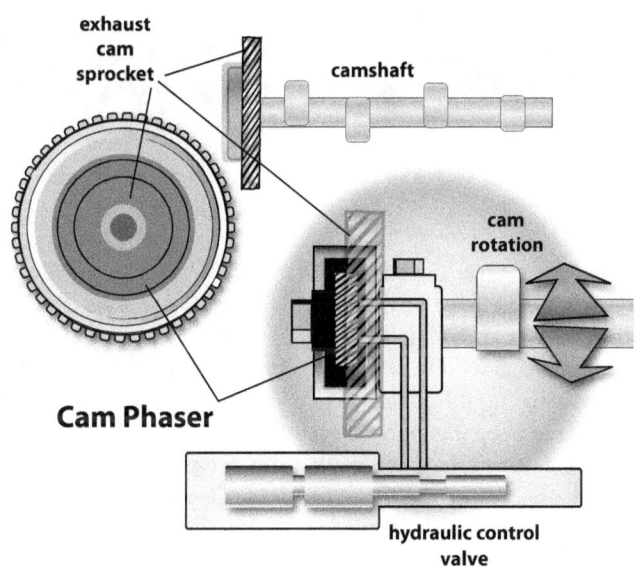

**Cam Phaser**

Some VVT systems change intake valve timing for better performance, and a few also control valve lift. The newest engines continually control both intake and exhaust valve timing to improve engine breathing over a wide rpm range.

# Emission Controls

## Ignition System Designs

While not generally classified as a dedicated emissions control device, the modern ignition system plays a vital role in improving engine performance and reducing harmful emissions.

Modern computer-controlled ignition systems generate the high secondary ignition firing voltage needed to promote good combustion in leaner air/fuel mixtures.

Once a firing solution is calculated by the computer, a voltage pulse is sent to a transistor, which then opens or closes the coil primary circuit switch-to-ground. The computer controls ignition firing frequency and timing, evaluating multiple sensors and choosing the best possible instant to fire each plug. Spark timing is constantly optimized for different engine speed, load, and temperature conditions.

**Caution:** Spark voltage is a health hazard. Modern coils generate heart- stopping voltages greater then 40,000 volts, and should always be treated with great caution.

Distributor caps and rotors have largely disappeared from production vehicles, but there are plenty left on the road.

- Alternatives include:
  **Coil-on-Plug ignitions with one coil per plug.** The coil is connected directly to the plug, eliminating the secondary ignition cable entirely. Each coil is individually triggered by the main powertrain computer. The coil switching transistor may be integral, or may be located inside the PCM or in a separate ignition module.
- **Waste Spark ignitions that fire two plugs per coil.** Also called DIS.
- **Coil-Near-Plug configurations** place a small coil near the plug, reducing the length of the secondary ignition wires.

DTCs commonly associated with ignition system failures:

**P030xx** - Misfire (random or cylinder-specific)

**P0350-P0362** - Ignition coil/circuit or control failures

## Common Misfire Causes

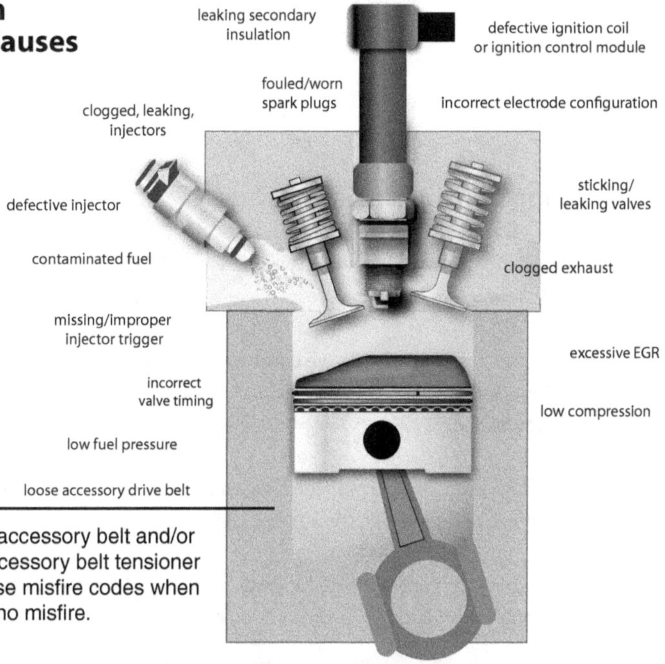

leaking secondary insulation

defective ignition coil or ignition control module

fouled/worn spark plugs

incorrect electrode configuration

clogged, leaking, injectors

defective injector

sticking/ leaking valves

contaminated fuel

clogged exhaust

missing/improper injector trigger

excessive EGR

incorrect valve timing

low compression

low fuel pressure

loose accessory drive belt

A loose accessory belt and/or weak accessory belt tensioner can cause misfire codes when there is no misfire.

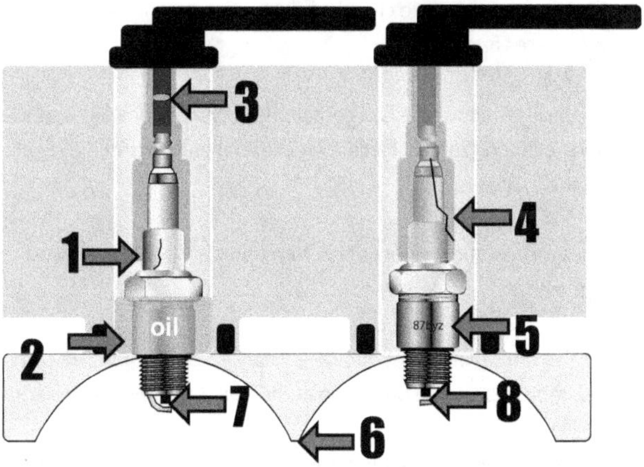

**1** **2** **3** **4** **5** **6** **7** **8**

oil

**Common ignition problems that result in misfire include (but are not limited to):**

**1)** cracked plug; **2)** oil in plug tube; **3)** open plug wire ; **4)** torn insulator; **5)** wrong heat range or improperly tightened plug; **6)** blown head gasket; **7)** worn plug with excessive gap; **8)** incorrect plug configuration (wrong plug design for the application).

## Emission Controls Review

- Powertrain Control Modules rely on oxygen sensor feedback to closely monitor and control the air/fuel ratio.

- Conventional oxygen sensors have only two useful states: rich or lean. They dither between rich and lean states to bracket the optimum air/fuel ratio, called *stoichiometry*.

- Wide range air/fuel sensors monitor the air/fuel ratio over a much wider range, and do not dither like conventional sensors.

- Catalytic converters operate efficiently when asked to clean the exhaust from an engine running at or near stoichiometry.

- Catalytic converters promote chemical changes that convert harmful gases into benign gases and water.

- Modern vehicles monitor the fuel containment system for leaks that allow fuel vapors to escape to the atmosphere. Several onboard leak-detection strategies are in use.

- Exhaust gas recirculation reduces NOx formation by limiting combustion temperatures.

- Variable valve timing increases engine efficiency, can replace a discrete EGR valve, and improves engine breathing at different engine speeds.

- Common ignition system failures cause engine misfire. Many are easily remedied by simple maintenance.

# TESTING EMISSIONS

# 8

# Testing Emissions

In this section we'll look at ways to test vehicles and use test results to make effective repairs. We'll also look at ways to prepare a vehicle for a tailpipe exhaust gas sampling test. In Chapter 9, we'll explain how to analyze your exhaust gas samples.

## Sniff, Fix, Test

Auto repair shops use **repair grade** analyzers to test exhaust gases. Analyzers display the amount of each measured gas as a percentage of the total sample, or in parts-per-million. Repair grade analyzers are used to diagnose and repair vehicle emission failures and driveability concerns. They are reasonably accurate, but less so than gas analyzers at emission test centers.

Currently, repair grade analyzers commonly measure five separate gases: Hydrocarbons (HC) and NOx, measured in parts per million (ppm), and CO, $CO_2$, and $O_2$, measured as a percentage of the gross exhaust sample.

Exhaust samples must not be diluted. Make sure that the exhaust system is leak-free and that the test probe is inserted 12-14 inches into the exhaust pipe.

Calibrate your gas bench regularly and replace the gas analyzer oxygen sensor and any specified filters at recommended intervals.

### Test Center Sniffers

The exhaust tester at the emissions test center is a lot larger, more accurate, and WAY more expensive. Test centers sample exhaust gases while the vehicle is driven on a special treadmill called a dynamometer. The "dyno" is a stress test that approximates real driving conditions. Driving the vehicle on the dyno gives a better indication of vehicle road emissions.

# Testing Emissions

Here are common steps for using your repair grade exhaust gas analyzer to test tailpipe emissions.

## Prepare the Vehicle

1. **Run the engine until it reaches normal operating temperature.** The thermostat should be open and the coolant temperature at least 180°F. Use a pyrometer and/or your scan tool to verify temperature.

2. **Verify that the engine is in closed loop:** Use your scan tool to determine fuel loop status. If the engine won't enter closed loop, find and fix the cause!

3. **Disable air injection:** Switch the air diverter valves to vent discharge air to the atmosphere, or disconnect and cap the air injection tubing. Do not **pinch off AIR hoses. You could damage the pump or hoses.**

4. **Insert the gas analyzer probe at least 14" into the tailpipe:** make sure the exhaust is leak-free.

5. **Turn all electrical accessories OFF.** This includes heating and air conditioning.

6. **Sample gases with the engine running at idle in Park or Neutral.** Test again (A/T only!) in Drive at idle. Then test again in Drive at a steady 1500-1800 rpm.

---

### Safety First when testing in DRIVE!

Keep your foot firmly on the brake, set the parking brake, and chock the drive wheels prior to any testing in gear! Consult manufacturer TSBs regarding possible damage from unacceptable loaded transmission test speeds and procedures.

Return the engine to idle in Park or Neutral as soon as analyzer readings stabilize.

**Caution: Some OEMs prohibit testing in drive since it can result in transmission damage. ALWAYS refer to OEM warnings before testing an unknown vehicle in drive.**

**Caution:** NOx testing at idle is ineffective since NOx increases radically with engine load. A road portable analyzer is your best bet here. If one is not available, listen for spark knock or ping, and/or monitor the knock sensor, if equipped. When you conquer ping, you knock out NOx! (See page 144 for NOx test tips.)

---

# Testing Emissions

## Catalyst Efficiency

In spite of improved engine efficiency, pre-catalyst exhaust still contains sizable concentrations of noxious and toxic gases. The converter's job is to clean the exhaust before it is pumped out the tailpipe to the atmosphere.

If the converter is doing its job, exhaust gas leaving the catalyst will be much cleaner than it was coming in. If we want to know the air/fuel ratio, we can simply look at the lambda number on our gas analyzer. Or we can do things the hard way and sample exhaust gases "upstream" of the converter.

Pre-cat samples allow us to evaluate the air/fuel ratio based on the uncatalyzed ratio of CO to $O_2$.

**Catalytic Converter**
**Oxidizes HC and CO -**
**Reduces NOx**

$CO_2$

$H_2O$

CO

NOx  HC

active materials
inside converter

**The Hard Way:** Sampling exhaust before it enters the catalytic converter is often easier said than done. You may have to remove a mixture access plug (now an endangered species) or unbolt the converter from the front pipe (noisy, dirty, ugly). Alternately, on some makes you can probe exhaust gas through an AIR injection pipe; at an EGR pipe; or at an exhaust gas backpressure hose attached to an EGR backpressure transducer.

Record all pre-catalyst readings and enter them into a lambda calculator that will spit out a simple lambda value.

**The Easy Way:** If your exhaust gas analyzer displays "Lambda" or "Air/Fuel Ratio," just use that and skip the rest.

---

iATN members have access to a free online lambda calculator. The iATN is the **International Automotive Technicians Network**. Membership is free (although supporting members have access to some very valuable features not available to non-paying members). At this writing, iATN links over 74,337 members from 155 countries, sharing over 1.7 million years of experience.

Check it out at ***http://www.iatn.net/***

## Lambda

To borrow an old slang phrase, **lambda** is "the bees knees." We love lambda. You will too. In fact we'll sing its praises repeatedly, probably until you tell us to shut up!

The lambda calculation displays the air/fuel ratio as a single number, based on exhaust gas measurements of $CO$, $CO_2$, $O_2$, and HC.

Lambda is sometimes called the *excess air factor* since it indicates how much **air** there is in the mixture **compared** to the amount of air needed for an ideal stoichiometric mixture.

---

- The symbol for **lambda** is the Greek letter $\lambda$.

- Lambda 1.00 represents an air/fuel ratio at the ideal, stoichiometric ratio of 14.7 to 1.

- Lambda lower than 1.0 indicates a **rich** mixture. Lambda 0.88 is **richer** than stoichiometry.

- Lambda is greater than 1.0 indicates a **lean** mixture. Lambda 1.20 is **leaner** than stoichiometry.

---

**Lambda values do not change from the effects of combustion or the chemical changes that occur inside the catalyst.**

**Lambda is an accurate reflection of the mixture inside the combustion chamber, even if there is no combustion!**

---

If your exhaust gas test equipment displays a lambda value, and you aren't using it, you are wasting its most valuable feature.

# Testing Emissions

## Lambda and Air/Fuel Ratio Conversion

**To calculate lambda from an air/fuel ratio,
divide the air value of the air/fuel ratio by 14.7.**

Examples:

The lambda value of a (rich) 11:1 air/fuel ratio is $\dfrac{11}{14.7} = .748\lambda$

The lambda value of a (lean) 16:1 air/fuel ratio is $\dfrac{16}{14.7} = 1.08\lambda$

---

**To calculate an air/fuel ratio from a known lambda value,
multiply lambda by 14.7.**

Examples:

The air/fuel ratio of a lean $1.2\lambda$ mixture is $(1.2 \times 14.7) = 17.64{:}1$

The air/fuel ratio of a rich $.89\lambda$ mixture is $(0.89 \times 14.7) = 13.08{:}1$

Note: The 14.7:1 stoichiometric ratio is for gasoline only.

For propane (LPG), use 15.87:1. For Methane (CNG) use 17.45:1

If you take nothing away from this book but an understanding of lambda—and then start using it—you are ahead of most people. Lambda can indicate an air/fuel imbalance or eliminate fuel and air as suspects—**immediately.**

Don't guess about air/fuel imbalances. Don't waste hours performing unnecessary tests. Use lambda and eliminate all doubt.

The Auto Emissions Bible

## Lambda Notes:

- **Fix all exhaust leaks ahead of the exhaust probe and disable the secondary air injection system or your lambda readings will not be accurate.** Air leaks dilute exhaust gas concentrations of CO, $CO_2$, HC, and NOx, and increase exhaust $O_2$. A 5% air leak increases exhaust oxygen by about 1.0% and makes the lambda value 5% leaner than it should be.

- **Lambda does not change due to the effects of combustion.** A tailpipe sample converted to lambda is as accurate as a lambda value calculated from exhaust gases ahead of the catalyst. (Sure makes tearing the exhaust apart look silly, doesn't it?)

- **NOx, oxygenated fuels, and octane differences do not have a significant impact on lambda calculations.**

---

If a vehicle fails an emissions test for high HC, but has a lambda value of 1.0, we know the hydrocarbon problem is **NOT** caused by a mixture imbalance.

On the other hand, if a vehicle fails for high HC with a lambda value of 1.08, we know the mixture is too lean. This can help us identify a lean misfire.

The beauty of lambda is that it tells us when the air/fuel ratio is a problem and, just as importantly, when it is **NOT** a problem.

**A Real World Example**

Here's a great example showing how lambda can identify the true nature of an air/fuel problem, and save you hours of wasted time spent fixing the wrong thing! A Chevy Venture shows up with a **P0171 DTC** (Fuel System Lean). An exhaust gas sample indicates a lambda value of **0.89**. If you've been paying attention, you know a lambda value below 1.0 indicates a rich condition. We have a rich condition and a lean code. How can that be?

Turns out a groggy oxygen sensor is sending a continuous low voltage (lean) signal. The PCM does not store a DTC for the bad O2 sensor. Instead it accepts the bogus input and thinks the system is lean.

It pays to double check the PCM's air/fuel diagnoses with lambda. (At Sam's shop, the **Lusty Wrench**, the emissions analyzer runs all day, every day.)

# Testing Emissions

## Mixture Testing Without Lambda

If you want to go to the trouble, you can use upstream CO and $O_2$ samples to evaluate the fuel mixture. This may be helpful if you have an old two- or three-gas analyzer that will not display lambda or give you enough data to calculate it manually. To evaluate the mixture using pre-catalyst CO and $O_2$ measurements, look for the following:

- **If the mixture is correct,** CO and $O_2$ should be **roughly equal,** and both should be lower than 2%.

- If CO is significantly higher than $O_2$, the mixture is rich.

- If $O_2$ is higher than CO, the mixture is lean.

- HC should be below 200 ppm.

- If HC and $O_2$ are both high, look for an ignition related misfire. If the ignition system tests good, with high HC and $O_2$, suspect a lean misfire condition.

- If HC and CO are both very high, the mixture is "pig rich." Something is dumping too much fuel into the cylinders. Look for a leaking injector, high fuel pressure, or an EVAP purge problem.

DTCs commonly associated with mixture control failures:

**P030xx** - Misfire

**P0171, P0174** - Mixture Lean

**P0172, P0175** - Mixture Rich

## Catalyst Efficiency

If the mixture entering the catalytic converter is close to stoichiometry (lambda between roughly 0.97 and 1.03), but tailpipe emissions fail to meet standards, the catalyst may be napping on the job!

$CO_2$ readings above 14.5% at idle indicate a working converter, but they don't indicate how well it's working under stressful conditions. And $CO_2$ won't tell us how well the catalyst is reducing NOx.

Here are two catalyst efficiency tests, one for **oxidation**, and one for **reduction** (on the next page).

### Oxygen Efficiency Test

This test evaluates catalyst oxygen storage capacity, an essential part of oxidizing CO and HC.

• Prepare the vehicle for testing.

• Run the engine until it reaches normal operating temperature.

• Temporarily disable AIR injection or pulse-AIR system (if equipped).

• Run the engine at a steady 1500-2000 rpm.

• Let gas analyzer readings stabilize. Look for $O_2$ to drop close to zero. If it does not, flow a small amount of propane into the air cleaner until exhaust $O_2$ drops to zero.

• *Snap* the throttle wide open *briefly* and let it snap closed again very quickly. Note the gas analyzer $O_2$ reading, just as its CO reading begins to rise. If the $O_2$ reading exceeds 1.2%, the converter has insufficient oxygen storage capacity and should be replaced.

---

If the O2 reading exceeds 1.2%, the converter has insufficient oxygen storage capacity and should be replaced.

---

# Testing Emissions

## Catalyst Reduction Test

NOx can be tested **only** with the engine under load, so you'll need a road portable NOx analyzer or a stationary NOx tester and dynamometer.

Use these steps to test catalyst reduction bed efficiency:
- Access the exhaust upstream of the converter.

- Drive the car on a repeatable course under moderate load. Record upstream NOx readings.

- Move the gas analyzer probe to the tailpipe. Repeat the test while monitoring the tailpipe NOx level under the same conditions. Record the downstream NOx readings.

- Divide the downstream NOx readings by the upstream NOx readings, then subtract the result from 1.00 to calculate catalyst efficiency in decimal form. (To convert your reading to a percentage, multiply the decimal number by 100.)

---

Example: The upstream NOx measurement is 1450 ppm; the downstream reading is 852.

852 divided by 1450 = 0.5875. Subtract 0.5875 from 1. (1 - 0.5875 = 0.4125).

Multiply 0.4125 by 100 to express the result as a percentage - 41.25%.

Compare the result to the test standards used in your local emissions test to determine if this vehicle will pass or fail.

---

New OEM catalyst reduction beds generally achieve 70-97% efficiency. Aftermarket converter suppliers must certify only a 30% minimum reduction bed efficiency.

CARB-approved replacement converters must certify either 60% or 70% efficiency in $NO_x$ reduction, depending on application.

---

DTCs commonly associated with low catalyst efficiency:

**P0420-P0422**
**P0430-P0432**   Low catalyst efficiency

# Testing Emissions

## Catalyst Efficiency Cranking Test

Here's a alternate catalyst oxidation bed efficiency test. This test tells us how much of a given amount of fuel can be oxidized by a hot catalyst. You'll need an emissions analyzer capable of measuring HC and $CO_2$ and a propane enrichment tool.

> **NOTE:** Please read through the entire sequence before starting. This test must be performed in a short period of time while the catalyst is hot, or it will not be accurate.

- Run the vehicle and your gas analyzer until both reach normal operating temperature.
- Hold engine speed at about 2000 rpm for at least two minutes to warm the catalytic converter to operating temperature.
- Turn the engine OFF.
- Complete the test within 5 minutes after turning off the engine or the catalyst will cool too much to work efficiently.
- **Disable fuel and ignition systems.**
- Attach a battery charger and crank the engine at **wide open throttle** for at least 10 seconds to clear all fuel from the intake and exhaust.
- Attach a propane source to a large, centrally located vacuum line (PCV or brake booster).
- Insert the gas analyzer test probe at least 14" into the tailpipe.
- Open the propane source valve; crank the engine for 15 seconds with the **throttle closed**.
- Read and record **peak** HC and $CO_2$.

Here are the test standards:

| If HC is: | then $CO_2$ should be greater than |
|---|---|
| 550 ppm | 1.7% |
| 750 ppm | 2.5% |
| 1000 ppm | 2.9% |
| 1250 ppm | 3.8% |
| 1500 ppm | 4.8% |
| 1750 ppm | 5.4% |
| 2000 ppm | 6.2% |
| 2250 ppm | 7.0% |

If the catalyst cannot pass the test, replace it.

# Testing Emissions

## Scan Tool Emissions Tests

Scan tool emissions inspections are not complicated or hard to understand. They use no emissions analyzers, dynamometers, or huge constant volume gas samplers. There are no exhaust gas test limits.

The idea behind scan tool tests is simple:

- **OBD II vehicles test themselves whenever they are driven. They store the test results.**

- Scan tool tests use that data and MIL status to determine if the vehicle has a problem that may increase harmful emissions to unacceptable levels.

If an emissions-related fault is detected, a Diagnostic Trouble Code (DTC) is stored in the computer and the computer turns on the Malfunction Indicator Light (MIL) in the dashboard.

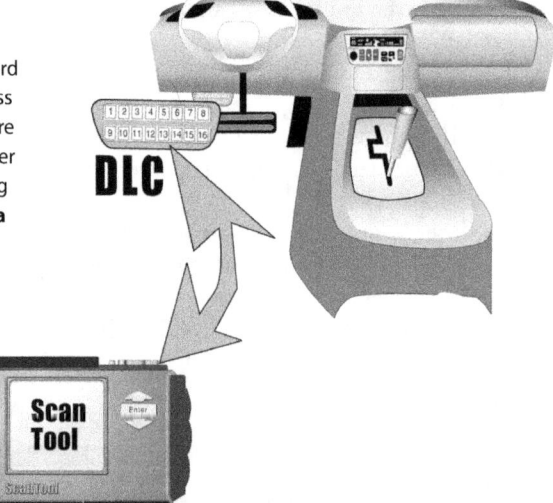

Scan tool tests rely on onboard monitoring test results to pass or fail a vehicle. These tests are performed on 1996 and newer OBD II vehicles by connecting a scan tool to the OBD II **Data Link Connector (DLC)** and scanning the vehicle for the following:

- **DTCs**
- **MIL commanded state**
- **Monitor status**

DLC

Scan Tool

---

Caution: The description of the scan tool emissions test that follows is a general overview, based on EPA suggested practice.

Check with your local emissions program for the exact standards being used in your area.

---

146

## Pass/Fail Test Standards

To receive a passing the grade, the vehicle must pass three important tests:

**MIL ON at**
**Bulb Check**

**1** The MIL must pass a simple "bulb check" visual inspection. The inspector turns the ignition to the ON position. The MIL should illuminate with the ignition ON to verify that the bulb works. (It may go out again on some vehicles if the key is left in the on position too long before the engine is started, but it must illuminate when the key is first switched ON, or fail the inspection.)

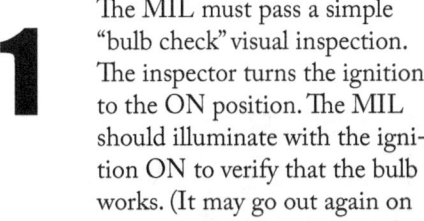

**MIL OFF**
**Engine Running**

**2** The MIL must go out when the engine is started—and stay out. The MIL should not come on with the engine running. If it does, the vehicle fails the inspection.

**3** A minimum number of non-continuous monitors **must be listed as complete** in the scan tool Readiness Status display. The software checks monitor completion status. Ideally, all monitors will be complete, but vehicles may still receive a passing grade on this part of the test under the following conditions:

- **1996-2000 model year vehicles may have a maximum of 2 incomplete non-continuous monitors.**

- **2001 model year and newer vehicles may have only 1 (some may allow 2) incomplete non-continuous monitor.**

> **Note:** Generally, a vehicle may pass with a DTC in memory **if** the PCM has **not currently requested the MIL**.
>
> This condition could indicate that the vehicle **had** previously stored a DTC, but that additional tests had run and passed three consecutive times, allowing the computer to turn the MIL off.
>
> The computer will eventually erase the DTC if the problem does not appear again during 40 engine warm-ups.

# Testing Emissions

## How Does a Vehicle Fail the Scan Tool Test?

The vehicle **fails** the test for any of the following:

**1** The MIL does not illuminate when the ignition is switched on.

**2** The MIL is ON with the engine running.

**MIL ON**

**3** The scan tool indicates that the PCM has requested the MIL ON, but the MIL is OFF.

**MIL OFF**
**Engine Running**

Test center technicians use the scan tool to check the MIL "commanded state." If the commanded state doesn't match the actual state, the vehicle fails. This identifies MIL circuits that don't work, and also prevents MIL circuit tampering.

OBD DTCs

**P0650** - MIL circuit fault

## Test, Don't Guess

In this chapter, we've put together a simple battery of tests that can identify common problems and their causes. Some require exhaust gas sampling and correct interpretation of test results. Others are as simple as looking at the MIL and hooking up a scan tool.

Use these tests to eliminate guesswork. They reduce the cost in time and unnecessary parts purchases associated with trial and error vehicle repair practices.

Do not use these tests only when you have a problem vehicle. If you never test a good vehicle, you'll never recognize a vehicle with problems when you see it. Use your gas analyzer regularly and these tests become a powerful diagnostic tool for identifying many problems, not just emissions failures.

Obviously, the tests you choose will depend on your tools and testers and your familiarity (and comfort) with their use. As the vehicle fleet ages, fewer vehicles will undergo tailpipe tests, with the OBD scan tool test dominating test procedures for the foreseeable future.

## Mini Quiz

Is this engine running richer or leaner than stoichiometry? (See page 139.)

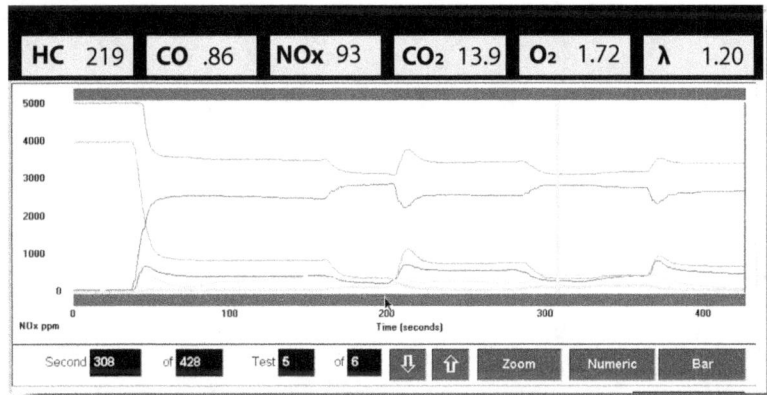

## Testing Emissions Review

- Repair grade analyzers are found in repair shops. Not as expensive or accurate as the analyzers at emissions test centers, they are still a valuable diagnostic tool when used properly.

- *Lambda* is the most accurate air/fuel ratio measurement. It may be calculated from exhaust gas readings, or displayed by repair grade analyzers with a lambda display.

- Lambda is not fooled by misfire.

- Catalyst efficiency can be measured using several tests.

- Scan tool emissions tests do not measure tailpipe gases. They use the status of the Malfunction Indicator Light and the stored results of onboard tests performed by OBD II software to determine if there are conditions that may result in increased harmful emissions from the tailpipe or fuel containment system.

# ANALYZING TEST RESULTS

8

## Analyzing Test Results

There is a big difference between conventional exhaust gas testing and OBD II scan tool tests. Both strategies have the same purpose (to get gross polluting vehicles repaired or off the road), but are unique and must be treated differently.

- **Traditional gas sampling** requires expensive equipment. Gas emissions tests sample the actual gases exiting the tailpipe, and may include a manual pressure test of the gas cap (but not necessarily the rest of the EVAP system). Some vehicles have been successfully "tweaked" to get them through a traditional tailpipe test. Tweaking has become increasingly difficult as fewer main engine controls are adjustable.

- **Scan Tool tests** are faster, cleaner, and simpler. They require less equipment and, since the vehicle is never driven on a dyno, they have a tendency to be less "scary" to vehicle owners.

- **Scan Tool tests** retrieve test results from an on-board system that monitors the vehicle during normal operation and stores codes for failures that *might* result in an increase in vehicle emissions, from exhaust gases or from vapor leaks to the atmosphere from the fuel containment system. We emphasize the word *might*, since the OBD II system does not measure emissions directly. It only tests components and systems that are responsible for emissions control to determine if they are working properly.

- The OBD II vehicle fails a scan tool test if the MIL is on KOER. To pass, the vehicle computer must be satisfied that a certain number of tests have run and that no faults have been detected in the process.

**Note:** Some states also perform a visual inspection with a floor mirror to verify the presence of the catalyst (or its shell!).

---

The diagnostic strategy for a **gas sample emissions failure** is to correct the cause for high emissions.

The diagnostic strategy for a **scan tool emissions test** is to correct a vehicle problem that causes the on-board system to store a DTC and turn on the Malfunction Indicator Lamp.

---

The Auto Emissions Bible

# Analyzing Test Results

## Maximize Your Efforts

While the main thrust of this book is to help you pass your vehicle emissions test, some cars have significant performance issues that can be remedied by using tailpipe and scan tool test results judiciously. Careful test analysis diagnoses vehicles that suffer from weak or uneven acceleration, poor fuel economy, stalling, misfire, hard starting, and a host of other issues, whether or not the vehicle is due for an emissions test.

**The same procedures already described for correcting excessive emissions problems will fix performance issues as well. Use the same procedures on a regular basis and emissions failures become routine; just another repair.**

Consider the following:

- **Low power and uneven acceleration** accompanied by large *positive* shifts in fuel trim can be caused by performance robbing problems like weak fuel pumps, low fuel pump voltage, and vacuum leaks. These same problems can also result in hard starting, especially when the engine is cold.

- **Poor fuel economy** accompanied by large *negative* shifts in fuel trim may indicate a leaking purge valve or failed fuel pressure regulator that is leaking raw fuel through its vacuum sensing port, or raising fuel pressure by restricting the fuel return line too much.

- **Misfire, hard starting hot or cold, poor transmission shift quality, and poor engine performance**, accompanied by large shifts in fuel trim may be traced to both leaking or severely clogged fuel injectors.

- **Frequent stalling** can often be traced to an EGR or purge valve that sticks open at idle.

- **Engine misfire** can result from clogged EGR passages that dump all the EGR output to a single cylinder.

- **Hard starting, poor idle control, reduced turbo boost, and multiple engine oil leaks** can all be traced on occasion to a faulty PCV system.

Get it? Test wisely. Document your data. Apply it as broadly as possible for maximum benefit, whether you are addressing an emissions failure or performance issue.

# Analyzing Test Results

## Interpreting Gases

How you interpret exhaust gas samples depends on the type of gas analyzer you're using. Some analyzers display 2,3,4, or even all 5 gases. Some display lambda value as well as gases.

Use the chart below to help you identify common causes for abnormal exhaust gas levels, and also to show how to use lambda to eliminate fuel control as a possible root cause for high emissions.

| GAS | Condition | Common Causes |
|---|---|---|
| HC | high | misfire; too rich or too lean; poor fuel atomization; possible EGR leak. If $O_2$ is also high, suspect ignition misfire or lean misfire |
| $O_2$ | high | too lean; unmetered air leak downstream of air-flow sensor; dead cylinder; exhaust leak; AIR system ON; inadequate fuel pressure/volume; $O_2$ sensor short to voltage |
| CO | high | too rich; excessive/unmetered fuel (don't overlook leaking pressure regulator or purge control valve); gas-diluted oil; $O_2$ sensor sticking low (dead sensor, exhaust leak near sensor, AIR system on at wrong time); major vacuum leak (speed/density system only); engine running too cold or faulty temp sensor; excessive fuel pressure |
| $CO_2$ | low | engine mechanical faults (pistons, cylinder walls, rings, camshaft, valves); low compression; incorrect valve timing; faulty catalytic converter; clogged exhaust |
| $NO_x$ | high | ignition timing too advanced; EGR faults; carbon buildup; cooling system faults; octane too low; stuck air intake preheat door; shift cable misadjusted; faulty knock sensor |
| **lambda** | **Condition** | **Causes** |
| high>1.03 | LEAN | lean mixture condition; see high $O_2$ |
| low<0.97 | RICH | rich mixture condition; see high CO |
| normal 0.97-1.03 | NORMAL | mixture is correct; look elsewhere for source of problems (engine mechanical, ignition, EGR, etc.) see low $CO_2$ or high $NO_x$ |

## Copy these pages and hang them on the wall!

## Using Lambda

Yes. We're back with lambda again. Four or five gas analysis is useful, but lambda nails an fuel/air ratio problem. End of story. Done deal.

Here's why: Lambda is not altered by combustion or by the catalyst. It doesn't give a hoot about misfire. Isn't fooled by catalyzed gases. Lambda tells us if our problem is or is not related to an incorrect air/fuel ratio, regardless of other factors, during and after combustion.

This chart compares the actual air/fuel ratio (lambda) to the oxygen sensor signal and system response. Based on the combination of these three elements, look in the last column for probable causes.

| Lambda - Actual A/F Ratio | PCM Response STFT + LTFT | $O_2$ Sensor mV | Things to check first |
|---|---|---|---|
| **(rich)** less than 0.97 | **positive-** greater than +10% | **(rich)** greater than 450mV | Incorrect PCM response to O2 sensor; check for codes, check PCM ground and power; PCM high bias voltage fault |
| **(lean)** greater than 1.03 | **positive-** greater than +10% | **(lean)** less than 450mV | Normal PCM response to lean condition Check for low fuel pressure; contaminated MAF; large vacuum leak; clogged injectors |
| rich or lean | **zero-** no fuel trim correction | rich or lean | Open loop; limp-in mode Check for codes, including transmission |
| **(rich)** less than 0.97 | **negative-** less than -10% | **(rich)** greater than 450mV | Normal PCM response to rich condition; Check for high fuel pressure; leaking injector(s); restricted exhaust (MAP system); EVAP purge fault |
| **(lean)** greater than 1.03 | **negative-** less than -10% | **(lean)** less than 450mV | Incorrect PCM response. PCM is receiving the correct $O_2$ signal but is still subtracting fuel; PCM fault; check for multiple codes; check PCM ground and power; check TSBs |
| **(lean)** greater than 1.03 | **negative-** less than -10% | **(rich)** greater than 450mV | $O_2$ sensor signal shorted to voltage, commonly to the sensor heater. |

When it comes to air fuel ratio issues, lambda is one-stop shopping. A single number tells us if we are rich or lean, or if we have an acceptable balance. Lambda—get it, use it.

## Notes

# TROUBLESHOOTING TIPS

**10**

# Troubleshooting Tips

## Fixing Cars

This is all about fixing cars; fixing them fast, fixing them right. In this chapter on troubleshooting, we're going to start with assorted tips, and bring it all together to give you what every emissions tech needs—a PLAN!

We have gathered common tests that will fix most cars quickly and efficiently. But there's a catch. You have to use the procedures in this section religiously, until they become second nature.

Later in this chapter, we'll look more closely at supplemental test procedures to add to your diagnostic arsenal.

This is not Rocket Science. Or is it Rocket Surgery? I can never remember.

## The VIR

Start with the VIR (Vehicle Inspection Report) issued by the test center. Use the VIR to identify the exact reason the vehicle failed.

The Vehicle Inspection Report should always be the starting point for diagnosis and repair of an emissions failure.

---

## Vehicle Inspection Report

Your vehicle emissions test results are shown below. If your vehicle passed, tear off the certificate and submit it with your registration documents. If your vehicle failed, it must be repaired and retested. Information on repair and retest requirements appears on the back of this certificate and on other documents provided for failed vehicles. Retain this certificate with your registration and other important automobile papers. It may be transferred to a new owner if you sell the vehicle.

| Test Fee | INDIVIDUAL TEST SUMMARY | | | | Final Result |
|---|---|---|---|---|---|
| | Tampering | Emissions | Evaporative System | Opacity | |
| 19.50 | PASS | FAIL | PASS | | FAIL |

| GENERAL INFORMATION | | | | |
|---|---|---|---|---|
| VIN | Vehicle Year | Make | Date | Time |
| | | | 04-FEB-2003 | 12:32:02 |

| INDIVIDUAL TEST RESULTS | Reading | Units | Limit | Result |
|---|---|---|---|---|
| Tampering: | | | | |
| Catalytic Converter | | | | PRESENT |
| Presence of Gas Cap | | | | PASS |
| Emissions ASM: | | | | |
| HC | 455 | PPM | 72.0 | FAIL |
| CO | 2.00 | % | 0.40 | FAIL |
| CO2 | 10.45 | % | | N/A |
| Evaporative System: | | | | |
| Pressurization of Gas Cap | 27.90 | In H2O | 6.00 | PASS |

---

# Troubleshooting Tips

## Baseline the Vehicle

Vehicle baselining is an important first step that records the exact vehicle condition *before* repairs are made. These conditions will be referenced after your repairs to confirm their success, or failure.

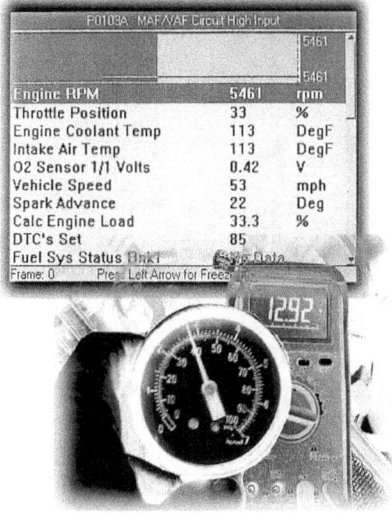

**Verify the test results on the VIR.** Make sure the reasons given for a vehicle failure are still present, and that they match the failure report details. This applies to both scan tool and tailpipe tests.

- **If the vehicle has failed a tailpipe test** performed on a dynamometer, but gas readings seem normal at idle when you test them, simulate the test conditions on a dyno, or on the open road with a portable gas analyzer. An alternate approach is to approximate the test load by placing the automatic transmission in Drive, firmly applying the brakes, and raising the engine speed to about 1300-1800 rpm. Remember, this is an approximation, used to detect emissions problems that appear only under load. **CAUTION:** Some OEMs strictly prohibit testing in Drive since it may result in very expensive transmission damage. Check OEM test guidelines. (Sorry to put the burden on you to check that, but it can't be avoided.)

- **If the vehicle fails a scan tool test**, check the MIL status, compare DTCs in memory to those recorded on the VIR, and look closely at freeze frame and datastream. Record all values before the repair for reference later.

Working on a pre-OBD II vehicle, and don't have access to datastream? Use a scope or graphing multimeter to monitor $O_2$ sensor activity going up a slight hill. Have an assistant drive while you look for abnormal $O_2$ sensor response indicating excessive fuel adjustments. Remember, a properly switching $O_2$ sensor is the best indicator of an engine in fuel control.

If the sensor does not switch rapidly enough, mixture control cannot be maintained. A new sensor may be all that's needed to get the vehicle to pass.

# Troubleshooting Tips

## Standard Test Procedures

If you properly baseline the vehicle, you may find your problem quickly. Many emissions problems are caused by a simple lack of maintenance. Baselining imposes discipline, and often uncovers basic problems.

- Check all vital fluids. Check battery voltage and condition.

- Start and warm the engine to normal operating temperature. Check system (charging) voltage. Correct battery and charging voltage issues first.

- Look for disconnected, bypassed, or damaged components. Not sure what emissions package you have? Look at the Vehicle Emissions Certification Information (VECI) label under the hood. All emission control devices will be listed there, often with component locations and vacuum hose routing schematics. Be careful: replacement hoods may have no label or the wrong label. When in doubt, check with the local dealership parts department by VIN. They may help you identify the calibration standard and components.

- Verify proper operation of the MIL (check engine light). It should illuminate KOEO for bulb check, but go out soon after the engine is started.

- Hook up your vacuum gauge and check engine vacuum using the steps on pages 174-175.

- Check for trouble codes. Review OBD II freeze frame and compare it to live data.

- If lambda and/or scan tool fuel trim values suggest a fuel delivery issue, test the fuel delivery system, including fuel pressure and volume.

- When you start fixing, troubleshoot and repair **component DTCs** before you tackle **system DTCs**. Component DTCs are often easier to diagnose and fix, and commonly cause system DTCs.

- Post repair, it may be necessary to clear all codes and retest, since some codes may alter the engine management strategies. There is no single hard and fast rule to guide us here. **Note:** OBD II Freeze Frame is erased when you erase Diagnostic Trouble Codes. Retrieve and save Freeze Frame data before erasing codes.

- Leave your scan tool connected and watch datastream. Almost all professional and many PC-based scan tools have a "movie" mode that lets you record and review datastream. Use graphing to view trends and look for voltage dropouts as you verify the success of your repairs.

# Troubleshooting Tips

## Check Basic Settings

- **Don't assume that ignition timing is correct.** Check and correct it, if necessary; use recommended procedures. Timing affects emissions.

- **If HC is too high, check the secondary ignition system, including available spark KV.** Check spark plug condition and gap. Make sure the correct plugs are installed. The wrong heat range spark plugs will cause misfire, increasing HC. Check plug pairs on DIS (waste spark ignition); gap erosion appearance depends on plug polarity.

- **If engine vacuum is abnormally low, find and correct the cause.** Don't overlook leaking EGR valves, especially at idle. Is the cam timing correct? A timing belt one tooth off is enough to cause problems. Is the exhaust restricted?

- **If the vehicle fails an emissions test for high CO, check the engine thermostat and ECT sensor.** An engine may warm up normally in your service bay, but surrender too much heat when driving on a cold day. If coolant temperature is too low, replace the thermostat before making other repairs. Engine coolant temperature affects injector ON-time. A lot.

- **Caution:** Engines operating in a "limp home" mode commonly run rich. In some cases, transmission codes may alter the computer operating strategy, causing increased emissions.

---

Incorrect camshaft timing or improperly adjusted valves can both cause low engine vacuum leading to DTC **P0108** -MAP circuit high

Incorrect ignition timing or other ignition system faults can lead to these DTCs: **P030xx** - Ignition misfire and **P0350-P0360** - Coil problems

High CO can be caused by faults associated with these DTCs:
**P0125-P0128** -ECT/circuit faults
**P0172, P0175** - Mixture Rich
**P07xx, P08xx, P09xx** - Transmission Faults

---

# Troubleshooting Tips

## Emissions Checklist

Not sure where to start? Like playing the odds? Then trust us, MOST emissions related problems start with one of the items on these two pages.

- **Fuel Supply:** fuel pump; fuel tank, fuel lines, hoses, and filters; fuel pressure regulator; fuel distribution; fuel quality.

- **EVAP System:** canister; vent lines/hoses; purge valve diaphragm; fuel cap; purge valves and solenoids; mechanical control systems; electrical control systems.

- **Fuel Metering:** mechanical control system; electronic control system; injector(s); throttle body; idle mixture control; cold start system; metering device; oxygen sensor and heater; engine coolant temperature sensor; air flow sensor; intake air temperature sensor; throttle position sensor; MAP or BARO sensor; CKP sensor; CMP sensor; knock sensor; VSS.

- **Idle Speed:** adjustment; idle air control valve.

- **Air Supply:** air filter; hot air intake system (TAC); intake manifold/gasket; vacuum/false air leak; turbocharger; supercharger.

- **Ignition System:** ignition module; primary wiring; coil windings or insulation; secondary ignition wires and boots; distributor cap and rotor; spark plugs; ignition timing; voltage drops in power or ground circuits.

- **Electrical/Electronic: PCM:** clear DTCs; check vehicle actuators; wiring (open circuit, high resistance, shorted); battery; charging system.

- **Exhaust Aftertreatment:** catalytic converter (empty, melted, damaged, low efficiency, or damaged downstream air tube plumbing); EGR (passages; mechanical control systems, electronic controls; valves/actuators); Secondary AIR injection system (belts, pumps, bypass and diverter, switches and valves; mechanical and electronic control systems, reed valves, check valves, other valves and plumbing)

- **PCV Valve:** include hoses and filters

# Troubleshooting Tips

## Emissions Checklist

- **Engine Mechanical:** internal short block; cylinder head structure/gasket; camshaft(s); timing belt, chain or sprockets; valves (lifters, rockers, dirty/burnt/bent/leaking); oil seals; valve adjustment; other seals and gaskets; VVT phaser and control system.

- **Engine Exhaust:** manifold and gaskets; check for high exhaust backpressure caused by obstruction/restriction. Don't just look for **high** backpressure: low restriction zoom-tube aftermarket exhaust systems can reduce backpresssure to a point where EGR will not function!

- **Engine Cooling:** fan; thermostat; radiators, coolers and caps; mechanical control systems; electronic control systems, antifreeze concentration; debris accumulations in radiator or condenser; low coolant level.

- **Vehicle Fluids:** coolant; crankcase oil; fuel

- **Transmission & Final Drive:** internal (mechanical/hydraulic); electronic control system; external controls (vacuum, cables and linkages); final drive ratio; tire size; excessive drag.

---

**Dear do-it-yourselfer!**
If you feel a little intimidated right now because you don't have ready access to professional grade test equipment, don't throw in the towel too soon.

Look at filters and fluid levels. Is the air filter so clogged that it could stop a 9 mm dead in its tracks? Is the coolant level so low that the coolant temp sensor is getting steamed but not boiled? Does the positive battery post have so much green goo growing on it that it looks like a herb garden? Are the accessory belts all properly tightened? Are the spark plugs so worn that you can stick your elbow into the gap?

There are plenty of simple things to check before you go looking for complicated and expensive solutions to basic problems. It's what skilled repair professionals do all the time.

Just sayin'. It's your money, dude.

---

# Troubleshooting Tips

## Basic Electrical System Tests

Check the vehicle electrical system.

- A faulty alternator or excessive RFI noise can cause the PCM to have a nervous breakdown. Check the battery and cables for integrity.

- Check for excessive AC ripple voltage; check that charging voltage is within specs; and make sure that all wiring harnesses and plug wires are routed *as originally installed*. Induced electrical noise in signal wires can really confuse the PCM.

- Charging system faults cause many PCMs to go into a "limp home" mode, usually accompanied by abnormally rich running.

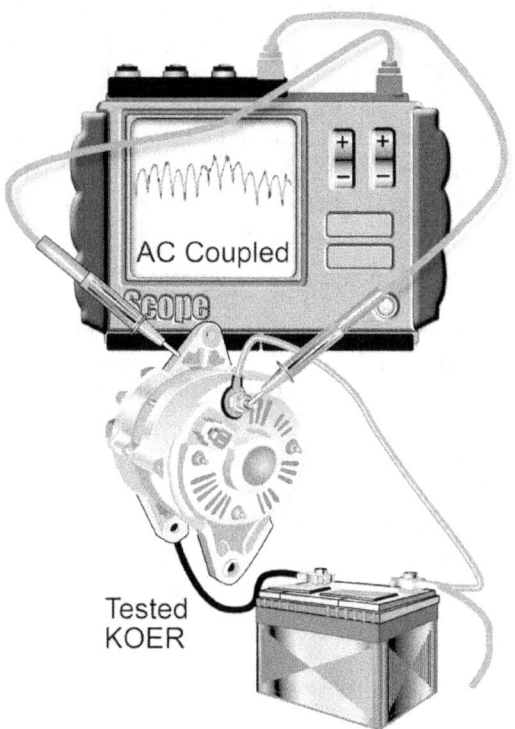

AC Coupled

Tested KOER

DTCs commonly associated with vehicle system voltage problems:

**P0560-P0564** - Battery and system voltage high, low, or out of range

An AC-coupled charging voltage waveform displays voltage spikes that may be misinterpreted by the computer as switching signals.

The Auto Emissions Bible

## Basic Electrical System Tests

Check the engine and chassis grounds. Many driveability and emissions problems result from loose or corroded ground connections.

Voltage drops in the main starting/charging system can cause more problems that we have room to list. Low charging system voltage can alter system performance and even shut the vehicle down completely when system voltage falls below a minimum level.

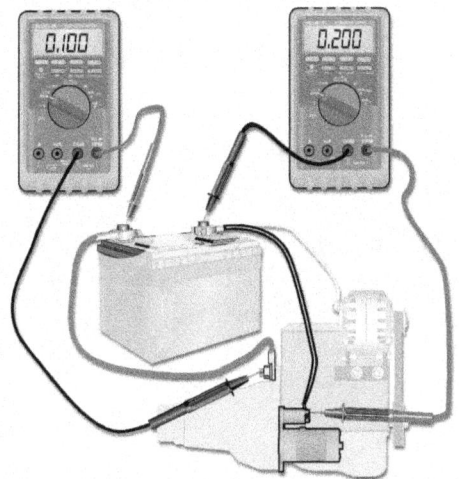

Use voltage drop tests to verify good connections in main cables, as shown in the image to the right.

Poor connections cause most electrical and many emissions problems. Electrical resistance lowers available voltage and skews sensor values.

Don't forget PCM power and ground voltage drop tests, especially when datastream values are not logical.

The graph to the right shows how the vehicle computer increases injector on-time if system voltage drops.

Some computers won't work at all below 10-10.5 volts.

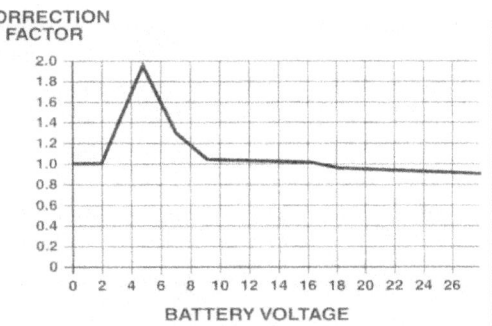

# Troubleshooting Tips

### Identify the Fuel System - Air Flow Style

How we interpret certain symptoms depends on the vehicle fuel system. Both MAF (mass air flow) and VAF (vane air flow) systems measure engine air intake mass (or volume) **ahead** of the throttle plate(s).

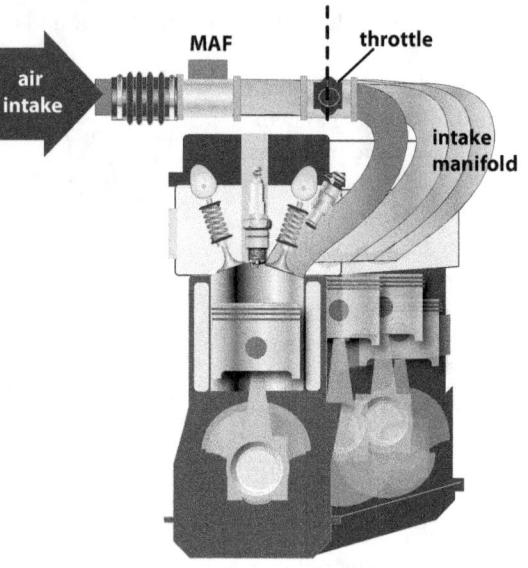

**MAF - Mass Air Flow** systems measure the **mass** of the air available for combustion. MAF measurements automatically compensate for air density fluctuations caused by changes in barometric pressure and air temperature. Mass Air Flow sensor signals may be analog or digital.

**VAF - Volume Air Flow** systems measure the **volume** of air available for combustion. Air mass (density) varies with temperature and barometric pressure, however. Denser, cooler air contains more oxygen for combustion. Intake Air Temperature (IAT) and Barometric Pressure (BARO) sensor inputs inform the PCM about these variables, so it can *calculate* mass.

- Mechanical problems that choke engine breathing also reduce air flow through air sensors, and that cuts fuel delivery.
- An engine intake manifold vacuum leak or a leak between the MAF/VAF and throttle plate reduces fuel delivery.
- Air that enters the engine without being reported to the PCM does not get its fair share of fuel.

Lean mixtures cause severe driveability problems and store DTCs for excessive fuel correction and/or lean misfires. These are among the most common DTCs resulting in vehicle emission test failures.

## Identify the Fuel System - Air Flow Style

mass air flow

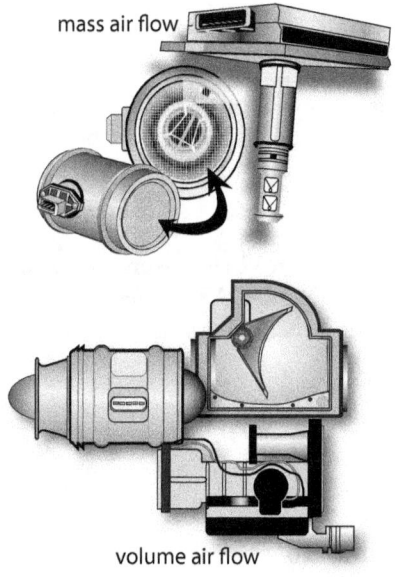

volume air flow

Air measurement meters come in all shapes and sizes; far too many to show here. Just remember that these meters are all located upstream of the throttle.

These DTCs are associated with air flow, MAP, and barometric pressure sensing components/circuits:

**P0171 or P0174** - (fuel system lean), or a manufacturer-specific DTC indicating excessive fuel trim corrections.

**P0100-P0104** - MAF faults \ **P0100** - VAF faults

**P0110-114** - IAT faults

**P0105-109** - MAP/BARO fault

**P0106-109** - BARO faults

# Troubleshooting Tips

## MAF Sensor Tips and Hints

Is a bad MAF causing your lean condition?

### Testing for a bad MAF

- **Remove the MAF and inspect the hot wire for signs of debris.** Hot wire mass air flow sensors can accumulate hairlike debris, as shown below. Some MAF sensing wires are hidden inside a sampling tube in the sensor body and are not visible.

- **Do not spray carburetor cleaner on MAF sensors and never immerse them in any cleaner.** Use a dedicated MAF cleaner. We have had good luck with CRC MAF cleaner. **Note:** Some sensors will not respond to cleaning.

- **We prefer not to use cotton swabs when cleaning MAF wires.** Swabs can leave tiny fibers that create new problems.

- **Let a MAF sensor cool at least 10 minutes before cleaning;** wait 10 minutes after cleaning before restarting the engine to prevent thermal shock to the sensor wire.

No, that isn't hair on the MAF hot wire, it's accumulated debris. Contamination is a common cause for inaccurate mass air flow sensor signals.

- **Install only a quality PCV valve from a known source.** Incorrectly calibrated PCV valves cause more problems than we can mention, including MAF sensing wire contamination. (See page 113.)

---

If a lean condition is causing a misfire, the OBD system may store a misfire DTC before it stores a lean mixture DTC. Freeze Frame fuel trim data will provide useful clues.

# Troubleshooting Tips

## MAF Sensor Tips and Hints

- **Install a premium air filter.** Cheap filter media do not trap enough dirt, and may even slough off tiny fibers that wrap themselves around the MAF sensing wire. Cheap filters are a false economy.

- **Compare LTFT for both banks in a V-engine.** If both banks are lean, the MAF is a possible cause. If one bank is lean but fuel trim in the other bank is normal, it's not likely that the MAF is at fault. (See page 184 for more on fuel trim.)

- **MAF sensors do not commonly store DTCs,** especially not for range or performance issues.

- **Look for a snap throttle peak voltage of at least 3.8 volts when testing an analog MAF.** Your digital meter, even set to MIN/MAX, may not be fast enough to record the peak voltage. We commonly look at an oscilloscope waveform when testing peak MAF voltage.

---

**Common Problem:**

Contaminated MAF sensors are chronic liars. They will commonly over-report air flow at idle, then under-report air flow at higher rpm. Add the two lies together and you get a unique problem.

Here's what happens:
At idle, the MAF reports **extra, phantom** oxygen that never reaches the cylinders. The PCM believes the lie and adds more fuel. Since there is less actual oxygen in the cylinders than what is reported, the extra fuel sends the air/fuel ratio rich. The oxygen sensor knows better. It sees low exhaust oxygen and reports the rich mixture. The PCM gets the message and shifts fuel trims *leaner*.

At higher rpm, the MAF makes its errors in the other direction. It under-reports true air mass. The PCM gets fooled again. Now it thinks there is less oxygen available than there really is, and it cuts back on fuel.

This can't possibly work: the vehicle is already running on a lean fuel trim when the fuel is cut back even further at higher rpm. The results are not pretty. Symptoms include hesitation, reduced power, and sometimes, the engine will even cackle like a chicken as it starves for fuel. Talk about laying an egg!

---

# Troubleshooting Tips

## Identify the Fuel System - MAP Style

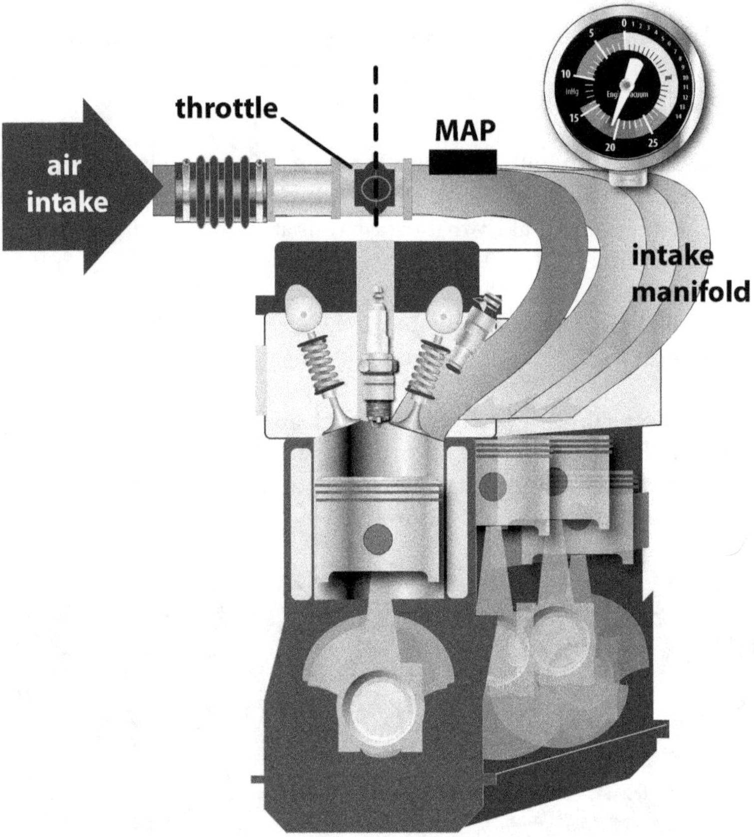

The term **Speed Density** refers to PCM fuel calculations based primarily on engine speed and air density. These two factors are the primary variables, although throttle position and intake air and coolant temperatures are also monitored. Oxygen sensor feedback fine tunes these calculations in closed loop.

These are also referred to as **Manifold Absolute Pressure** systems since they measure manifold pressure as (a big) part of their fuel delivery calculations. Manifold pressure normally increases as the throttle is opened.

MAP systems measure manifold pressure **below** the throttle plate. Compare this to MAF and VAF systems that measure air mass **before** the throttle.

The Auto Emissions Bible

# Troubleshooting Tips

## MAP (aka Speed Density)

MAP systems are robust. You won't get wealthy selling replacement MAP sensors. If the system has a weakness it's that it doesn't make adjustments for engine wear or any other factor that might affect engine breathing. This can skew its calculations over time. It is a major reason vehicle makers have largely shifted to MAF systems, since (ideally) the mass of all air entering the engine is measured. No further adjustments are necessary when calculating the oxygen available for combustion.

### MAP Troubleshooter

MAP sensors are very reliable. If you have a MAP sensor DTC, suspect the wiring or vacuum hose before you blame the sensor itself.

Expect to see decreased IAC (Idle Air Control) counts in a MAP system with a manifold leak to atmosphere. Engine speed increases due to added air from the leaks plus increased injector pulse width. The PCM sees this and attempts to lower the idle by closing IAC (Idle Air Control motor). Decreased IAC counts are a common symptom of a leak.

*Where* a leak occurs has a huge effect on the DTC or driveability symptom it causes. While a centralized manifold air leak may simply cause no change or an increased idle with the IAC closed, a misfire code can be stored if a leak is localized at a single cylinder.

> **Caution:** Don't just assume that the MAP PID is correct. If you have any doubt about the accuracy of the MAP data, take an actual vacuum reading at the intake manifold using a quality vacuum gauge.
>
> Compare the reading to the MAP PID. With a sensor as critical as MAP, sometimes the extra effort needed to cross reference our tests pays off. Remember, to calculate MAP, subtract engine vacuum from barometric pressure.

# Troubleshooting Tips

## MAP Sensor As Diagnostic Tool

Some vehicles open the EGR valve during closed-throttle deceleration to test EGR flow.

Here we see rpm drop during a closed throttle deceleration. Intake manifold pressure is very low (vacuum is high). As the EGR opens, we can see MAP voltage increase, indicating an increase in pressure as exhaust gas enters the manifold.

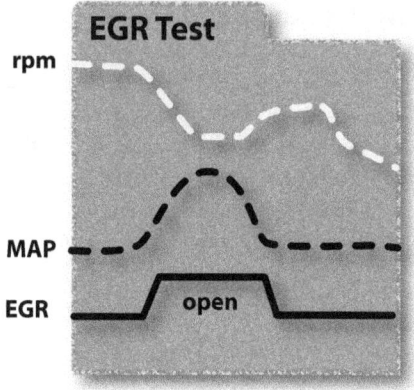

Since the engine is "coasting" at this time, there is no adverse affect on performance. If the MAP voltage does not increase enough to indicate good flow, the test fails, and may light the MIL. This usually means that the EGR valve is not opening, or that the EGR exhaust path is clogged.

We have added a ring of numbers around the outside of a standard vacuum gauge. The numbers printed on the gauge face indicate vacuum; our numbers indicate absolute **pressure**. As the throttle opens, the pressure of the atmosphere surrounding the vehicle rushes into an area of LOWER pressure inside the intake manifold. Pressure inside the manifold increases.

It is more logical to think of manifold **pressure** than to think of **vacuum**.

It helps us understand turbo and superchargers, and why exhaust gas flows into the intake manifold when the EGR valve opens.

## Backpressure Tests

High exhaust backpressure causes many performance problems that are often misdiagnosed. Plugged exhausts foul up all sorts of things and cause excessive fuel corrections, misfire, and a host of other problems.

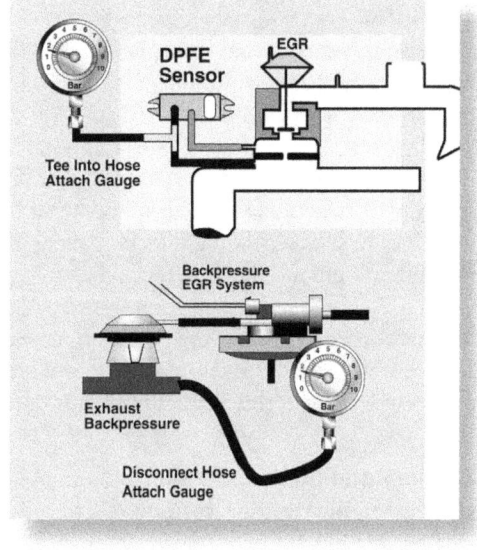

Testing exhaust backpressure isn't always the easiest thing to do, however. On makes with "friendly" exhaust configurations, you can unscrew an easily accessible $O_2$ sensor and insert a backpressure gauge in its place. With other configurations, you may have no choice but to drill test holes in the exhaust, take your backpressure reading though the test hole, and plug the holes you made when you're done. Ugly.

Vehicles with backpressure sensing EGR systems make testing backpressure easier. They come with built-in pressure sensing hoses that modulate EGR flow to match engine load. So for test purposes, all we need to do is remove the backpressure hose and connect our pressure gauge. Neat.

In Ford's DPFE (Differential Pressure Feedback EGR), Chrysler's backpressure sensing EGR systems, and many Asian vehicles, a small heat resistant hose connects an exhaust port and an EGR pressure-sensing device. Connect your backpressure gauge as shown in our illustration.

Don't be timid: raise the engine speed as you monitor backpressure. A reading over 1 psi at 2000 rpm is suspect and requires further investigation.

# Troubleshooting Tips

## The i Vac

We have this theory that if Apple® re-introduced the humble vacuum gauge to the automotive market as the iVac, people might take it seriously.

Okay, so it won't take pictures or play your favorite music, but think about it this way: If the engine is an air pump, what better tool could there be to test its ability to suck air than a vacuum gauge? And compared to most automotive test equipment, the iVac is cheap.

The hardest part of using the "iVac" on newer engines may be finding an easy, centrally located vacuum port on the intake manifold. But once it's connected, you can run a set of tests in minutes that will tell you volumes!

### Engine Running Vacuum Test

Use the **engine running vacuum test** to provide an overall picture of how well the engine is working as an air pump.

Attach a vacuum gauge to a centrally located manifold vacuum port.

- Run the engine at idle. A steady reading of 18-20 inHg is a sign of overall engine mechanical health. Gradually raise the engine speed to about 2000 rpm and look for a small increase in vacuum.

Normal and Steady

**Note:** In some engines, idle vacuum of 16-18 inHg may be normal. This is true of engines using multiple valves per cylinder, or vehicles with hi-flow exhausts.

- Snap the throttle closed and look for high momentary vacuum readings of 25 inHg, or more.

This test tells us about cylinder sealing.

Throttle
Snap Closed

# Troubleshooting Tips

Snap Throttle

- Now briefly snap the throttle wide open: vacuum should drop to zero briefly before returning to normal. If it sucks, its good!

Needle Bounces - Misfire

- A bouncing needle usually indicates misfire, burned valves, or weak valve springs.

Restricted Exhaust

- Low overall vacuum may indicate cam or ignition timing faults, low compression, leaking EGR, or a clogged exhaust.

If vacuum falls gradually at steady rpm, check for a clogged exhaust that accumulates exhaust backpressure, reducing intake vacuum.

# Troubleshooting Tips

## The Importance of Oxygen Sensors

Oxygen sensors are the heart of vehicle emission controls.

Here are examples of the $O_2$ sensor's many responsibilities:

- **Catalyst Efficiency Tests** - Pre- and post-catalyst oxygen sensor voltages are compared to measure catalyst efficiency, indicating how well the catalyst stores oxygen.

- **Misfire Detection** - The primary misfire detection input is the crankshaft speed sensor, but the oxygen sensor is commonly used to verify high exhaust oxygen levels resulting from cylinder misfire.

- **Secondary AIR Injection** - The oxygen sensor expects to see increased exhaust oxygen levels to verify air pump operation.

- **EGR Flow Testing** - If the EGR is working properly, exhaust oxygen levels should decrease in speed density engines as the EGR flow increases.

- **EVAP Purge** - Note: exhaust oxygen may go up or down during purge, depending on the contents of the canister. Bottom line: there should be *some* change.

| Test ID (TID) | Component ID (CID) | Test Value | Min Limit | Max Limit | Units |
|---|---|---|---|---|---|
| $11: O2 Sensor Monitor | $80: B1S1 Half-Cycle Counter | 47.000 | 26.000 | | count |
| $13: O2 Sensor Monitor | $80: B1S1 Big Slope Counter | 255.000 | 10.000 | | count |
| $19: O2 Heater Monitor | $80: B1S1 Hot Trend Counter | 255.000 | 6.000 | | count |
| $1A: O2 Heater Monitor | $80: B1S2 Hot Trend Counter | 255.000 | 6.000 | | count |
| $1B: O2 Heater Monitor | $80: B1S1 Heater Voltage | 200.000 | 180.000 | | mv |
| $1C: O2 Heater Monitor | $80: B1S2 Heater Voltage | 200.000 | 180.000 | | mv |
| $21: Catalyst System Monitor | $00: Bank 1 Switch Ratio Frequency | 0.000 | | 56.160 | % |
| $22: Catalyst System Monitor | $00: Bank 2 Switch Ratio Frequency | 2.340 | | 56.160 | % |
| $31: O2 Sensor Monitor | $80: B2S1 Half-Cycle Counter | 255.000 | 26.000 | | count |
| $33: O2 Sensor Monitor | $80: B2/S1 Big Slope Counter | 86.000 | 10.000 | | count |
| $39: O2 Heater Monitor | $80: B2S1 Hot Trend Counter | 255.000 | 6.000 | | count |
| $3A: O2 Heater Monitor | $80: B2S2 Hot Trend Counter | 255.000 | 6.000 | | count |
| $3B: O2 Heater Monitor | $80: B2S1 Heater Voltage | 220.000 | 180.000 | | mv |
| $3C: O2 Heater Monitor | $80: B2S2 Heater Voltage | 220.000 | 180.000 | | mv |

OBD II vehicles test the oxygen sensors, displaying test results in Modes 5 or 6. This Mode 6 capture shows tests and test results. (**See Chapter 12: Mode 6**). Not all vehicles display oxygen sensor test data. Clearly, oxygen sensors are pivotal players, used for closed loop fuel control and as a part of several OBD II monitors.

The Auto Emissions Bible

# Troubleshooting Tips

## Oxygen Sensor Quick Tips

Here are two quick $O_2$ sensor tests to verify the response and voltage range of traditional zirconia oxygen sensors. You can measure the sensor voltage response with a DMM, but an oscilloscope displays signal voltage transitions more accurately and in greater detail.

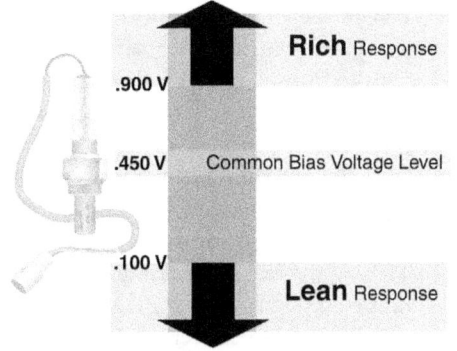

- **Force the system lean** while monitoring $O_2$ sensor voltage. Unplug one injector. Without combustion, dead cylinders turn into air compressors. The added blast of oxygen should cause $O_2$ sensor voltage to drop. (If the injectors are inaccessible, you may try to force the system lean by disconnecting a large, centrally located vacuum hose, such as the brake booster hose or a PCV hose. Be careful, however, because MAP sensor systems may interpret the sudden drop in vacuum as an increase in engine load, and compensate by throwing the system into maximum fuel enrichment.)

**What to expect:** Look for $O_2$ sensor voltage to fall like a stone, below 100 mV. The transition to a low voltage should be abrupt. If the voltage drops slowly, the sensor is lazy and needs to be replaced.

- **Force the system rich** while monitoring $O_2$ sensor voltage. Begin to add a small flow of propane into the air intake duct. Gradually increase the flow.

**What to expect:** $O_2$ sensor voltage should rise. When you reach a reading of 800 mV, suddenly increase the flow rate. Look for a rapid increase in sensor voltage, above 900 mV, just before the engine floods.

Sensor voltage should increase slowly at first as the fuel system tries to compensate for the gradual addition of HC from the propane. If the sensor is good, however, that last blast of propane should cause a **rapid increase** in sensor voltage.

# Troubleshooting Tips

## Oxygen Sensor Tips

- **Oxygen sensors are diagnostic sensors.** In addition to its role in the engine fuel control strategy and for catalyst monitoring, the oxygen sensor may also be used to check the AIR system, EGR flow, EVAP purge, and function as a cross reference for the misfire monitor.

- **If sensors and their heaters work properly, $O_2$ signal voltage should begin to fluctuate within a very short time after engine start-up.** This is referred to as a **time-to-activity.** You can see cold sensors do this by watching the oxygen sensor signal voltage right after you turn the ignition on. A cold oxygen sensor PID should show bias voltage cold. KOEO voltage should then decrease from bias voltage to a lean reading (low voltage) as soon as the heater gets the sensor hot enough to work. Remember, do this test KOEO, and be ready to monitor the sensor quickly. Some sensors wake up in less than a minute!

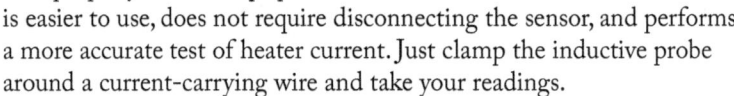

- **Do not test oxygen sensor heater elements with an ohmmeter.** A heater that falls within an acceptable resistance range may still not heat properly. A low amps probe is easier to use, does not require disconnecting the sensor, and performs a more accurate test of heater current. Just clamp the inductive probe around a current-carrying wire and take your readings.

- **GM $O_2$ sensors are tested to see if the heaters get them hot within a specified time.** While there may be a few early oddball cars without them, expect OBD II oxygen sensors to have electric heaters. HO2S means Heated Oxygen Sensor. Again: Do not trust heater element resistance tests. Put the ohmmeter away and measure current.

- **Check wiring and sensor harness connectors carefully.** Wiring damage or high connector resistance are common problems, especially with downstream sensors that are located beneath the car at the rear of the catalyst. This undesirable location exposes them to the effects of water, heat, and road salt.

# Troubleshooting Tips

## Oxygen Sensor Tips

- **Poor exhaust system grounds can skew oxygen sensor voltage.** Some sensor problems have been corrected by simply loosening and re-tightening the exhaust manifold fasteners! The oxygen sensor heater may have a dedicated ground circuit; check it separately when a heater won't work. A high resistance ground reduces the voltage dropped across the heater element, which also reduces heater output.

- **False air can skew oxygen sensor voltage.** Exhaust air leaks ahead of the oxygen sensor dilute the exhaust, adding excess oxygen. This fools the PCM into thinking the mixture is leaner than it is. Repair exhaust leaks.

- **Oxygen sensor bias voltage of 0.5 volt is common— but not universal—** so check the spec on unfamiliar vehicles. Some $O_2$ sensor circuits have no bias voltage.

---

There are over 100 generic OBD DTCs related to oxygen sensors and heaters - too many to list here.

Check in these ranges:

**P0030-P0064**
or
**P0130-P0167**
or
**P2195-P2198**
or
**P2231-P2256**

---

# Troubleshooting Tips

## Oxygen Sensor Tips

- **Initial scope setting for O2 sensors** —1 second/DIV and 200mV/DIV. Run the engine at 2000 RPM. Readjust the scope to 200ms/DIV and 200mV/DIV to zoom in for a closer look at the signal. (Use a meter or scope with at least a 10 megohm impedance.)

- **Normal zirconia oxygen sensor frequency at 2000-2500 RPM is 0.8-2.0 Hertz (cycles/second).** If oxygen sensor frequency is greater than this, look for a misfire condition.

- **Excessive oxygen sensor frequency normally indicates a misfire condition.** Hash indicates blasts of oxygen inside the exhaust from misfiring cylinders.

- **If a conventional zirconia oxygen sensor signal voltage is greater than one volt**—and stays high all the time—check for a voltage crossfeed between the sensor heater voltage supply and signal wire. This is more common than you think! Don't use this rule of thumb to condemn a wide range air fuel sensor! (See pages 108-109.)

- **If the bias voltage is pegged at 0.5 volt all the time, suspect an open circuit, a contaminated sensor that cannot respond, or a faulty heater that doesn't bring the sensor to operating temperature.** Heater operation is especially important for sensors mounted farther away from the engine and for post-catalyst sensors exposed to cooler exhaust gases.

- **If the oxygen sensor voltage is stuck at zero,** unplug the sensor while monitoring the sensor voltage on your scan tool. If the voltage increases to bias voltage as the sensor is unplugged, the sensor is shorting the circuit to ground.

    **Caution:** Some oxygen sensor circuits have no bias voltage. Check the vehicle specs.

The Auto Emissions Bible

# Troubleshooting Tips

## Oxygen Sensor Tips

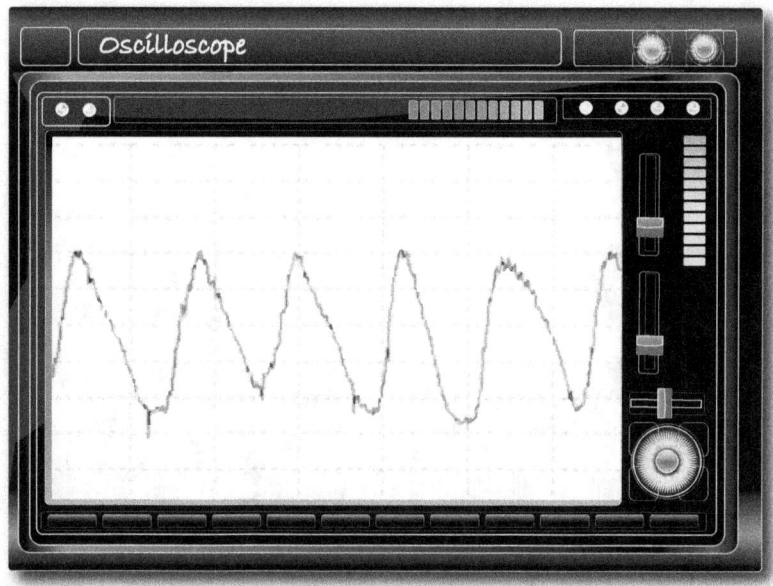

The voltage waveform of a zirconia sensor is displayed on an oscilloscope. Note how the waveform voltage pattern rises and falls regularly to indicate the sensor's response to alternating rich and lean fuel system corrections. This switching overshoots slightly in both directions; the "dithering" brackets the desired air/fuel ratio. This indicates that the fuel system is in control.

Some OEMs power the oxygen sensor heater through the PCM. Make sure that any replacement sensor has a heater with the same wattage as the original or you may be the proud owner of a fried PCM.

# Troubleshooting Tips

## Oxygen Sensors and Catalyst Efficiency

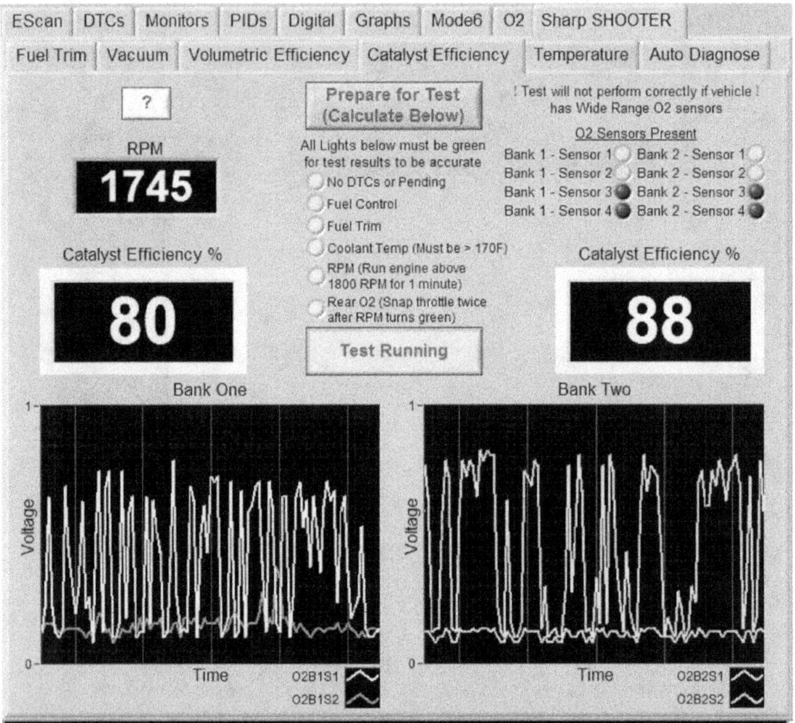

| EScan | DTCs | Monitors | PIDs | Digital | Graphs | Mode6 | O2 | Sharp SHOOTER |

| Fuel Trim | Vacuum | Volumetric Efficiency | Catalyst Efficiency | Temperature | Auto Diagnose |

**?**

**Prepare for Test**
**(Calculate Below)**

! Test will not perform correctly if vehicle !
has Wide Range O2 sensors

**RPM**

**1745**

All Lights below must be green
for test results to be accurate

○ No DTCs or Pending
○ Fuel Control
○ Fuel Trim
○ Coolant Temp (Must be > 170F)
○ RPM (Run engine above 1800 RPM for 1 minute)
○ Rear O2 (Snap throttle twice after RPM turns green)

O2 Sensors Present

Bank 1 - Sensor 1 ○    Bank 2 - Sensor 1 ○
Bank 1 - Sensor 2 ○    Bank 2 - Sensor 2 ○
Bank 1 - Sensor 3 ●    Bank 2 - Sensor 3 ●
Bank 1 - Sensor 4 ●    Bank 2 - Sensor 4 ●

Catalyst Efficiency %

**80**

**Test Running**

Catalyst Efficiency %

**88**

Bank One

Voltage — Time — 02B1S1 / 02B1S2

Bank Two

Voltage — Time — 02B2S1 / 02B2S2

Screens from an ESCAN scan tool show superimposed voltage waveforms of up-stream and downstream oxygen sensors in banks one and two of a v-engine. Sharp up and down signals are made by the upstream sensors as the fuel system dithers back and forth between rich and lean. The flatter trace across the bottom is made by the downstream sensor. Efficiency numbers are calculated for both cats. This vehicle's OBD II monitor passed these cats. Test standards vary by year, make, and model.

The vehicle computer monitors catalyst efficiency. It compares voltage signals from an oxygen sensor located at the catalyst inlet to signals from a second sensor located at the catalyst outlet.

The inlet sensor shows oxygen swings in the untreated gases coming straight from the engine exhaust ports.

If the catalyst is working, the signal from the downstream sensor will be flatter, since the exhaust has been cleaned.

The Auto Emissions Bible

## Oxygen Sensors and Catalyst Efficiency

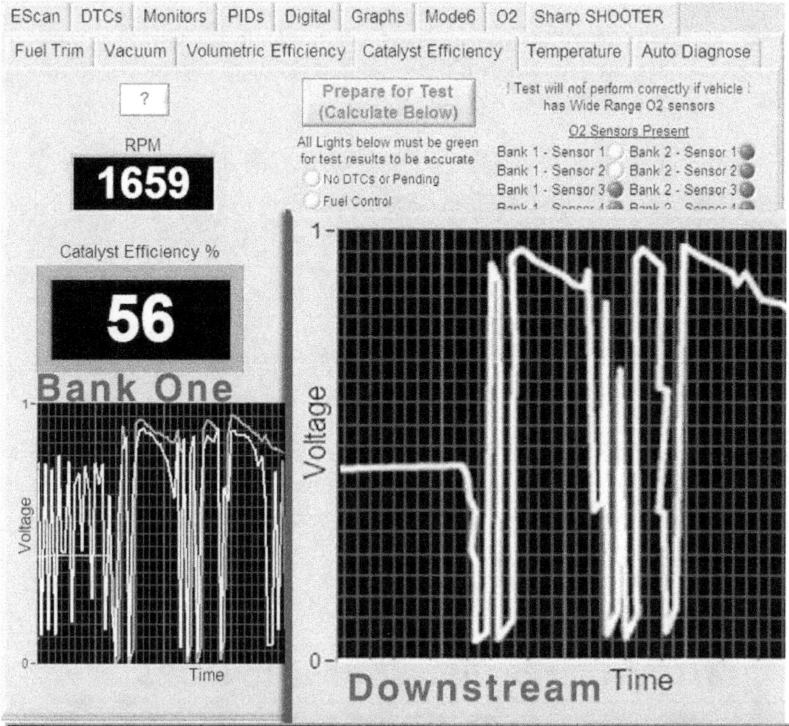

| EScan | DTCs | Monitors | PIDs | Digital | Graphs | Mode6 | O2 | Sharp SHOOTER |

| Fuel Trim | Vacuum | Volumetric Efficiency | Catalyst Efficiency | Temperature | Auto Diagnose |

?

Prepare for Test
(Calculate Below)

! Test will not perform correctly if vehicle !
has Wide Range O2 sensors

RPM

**1659**

All Lights below must be green
for test results to be accurate
◯ No DTCs or Pending
◯ Fuel Control

O2 Sensors Present
Bank 1 - Sensor 1 ◯  Bank 2 - Sensor 1 ●
Bank 1 - Sensor 2 ◯  Bank 2 - Sensor 2 ●
Bank 1 - Sensor 3 ●  Bank 2 - Sensor 3 ●
Bank 1 - Sensor 4 ●  Bank 2 - Sensor 4 ●

Catalyst Efficiency %

**56**

Bank One

Voltage

Time

Voltage

Downstream Time

We pulled out the downstream waveform and showed it separately in an enlarged
window. See how the downstream sensor is starting to mimic the upstream sensor?

This catalyst has issues. Its efficiency has dropped to 56 percent. It has illuminated the MIL and stored a low catalyst efficiency code. The larger screen to the right is an enlarged view of the downstream sensor. See how its sharp up and down trace and amplitude are starting to mirror the waveform from the front oxygen sensor?

The exhaust is not being oxidized as it passes through the catalyst. Eventually, if the catalyst stops working altogether, the upstream and downstream sensor waveforms will be identical.

# Troubleshooting Tips

## Using Fuel Trim Data for Accurate Diagnosis

Long- and short-term fuel trim data are a great diagnostic tool. A power user tool. Fuel trim data tell us if the computer needs to add fuel or subtract fuel to maintain the desired air/fuel ratio.

Fuel trim values can be viewed on a scan tool. In GM data, fuel trim numbers greater than 128 indicate that fuel is being added: fuel trim numbers below 128 indicate that fuel is being subtracted.

OBD II generic datastream indicates fuel trim changes as a *percentage* change from the base fuel schedule (zero percent). **Positive** percentage values indicate that fuel is being **added; negative** percentages, that it is being **subtracted**.

 In general, LTFT + STFT should be less than **±10%**. If LTFT + STFT exceed 20% positive or negative, there is a problem. The size of the correction needed to store a DTC varies. Check vehicle specs to be sure. Test known good cars by type and record the results for future reference.

## Short Term
### fast correction in response to HO2S

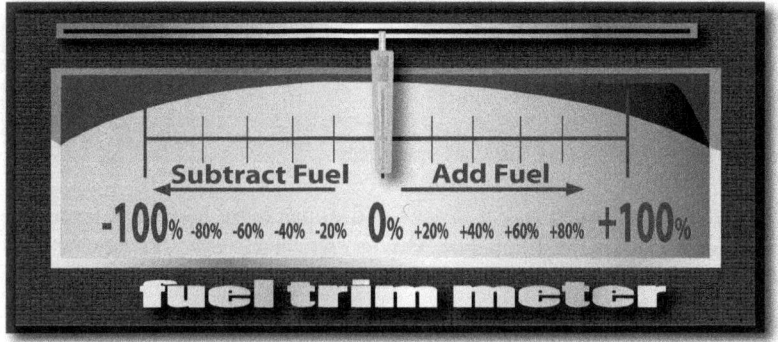

Subtract Fuel · Add Fuel

-100% -80% -60% -40% -20% 0% +20% +40% +60% +80% +100%

fuel trim meter

## Long Term
### slower correction based on Short Trim

Fuel trim is extremely effective for tracking fuel system problems because the PCM tracks and records fuel correction all the time. Then it stores long term trends in memory.

The Auto Emissions Bible

## Using Fuel Trim Data for Accurate Diagnosis

- **Scenario:** A MAF system shows LTFT plus STFT over +10% at idle.
  The PCM is adding fuel at idle, suggesting a fuel supply issue: faulty fuel
  pump, clogged fuel filter, etc. But before you fuss with fuel pressure tests,
  try this: Raise the engine speed to 3000 rpm and watch fuel trim again.

If **STFT plus LTFT** shifts back towards zero, you know there's enough fuel. Dispense with fuel pressure and volume tests, for now. Look instead for too much air at idle, possibly caused by a medium-sized vacuum leak, or improper cam timing.

**+15%**
*at idle*

**STFT**

**-3%**
*rpm raised*

- **Scenario:** A speed density system reads rich at idle (LTFT plus STFT are greater than 10% negative). Suppose raising rpm to 3000 brings fuel trim very close to zero. You may have just found a leaking EGR valve. (EGR is a much smaller part of total intake at 3000 rpm.)

# Troubleshooting Tips

## Fuel Supply Testing

If fuel pressure at the injectors varies from specifications, the vehicle may run richer or leaner than intended.

Use lambda to confirm the *true* air/fuel ratio.

Check lambda at different loads and engine speeds. If your numbers are consistently rich, abnormally high fuel pressure may be to blame. A faulty pressure regulator or a restricted return line could be at fault.

- If pressures are normal, don't overlook the possibility of fuel being sucked into the engine from a **faulty evaporative purge valve**.

- If the system is **lean under all conditions**, or trends lean under high fuel demand conditions, check fuel pressure and volume and pump voltage.

**injector pulse widths equal - only fuel pressure changes**

low pressure | **normal pressure** | high pressure

if normal fuel pressure is 45 psi

not enough fuel | **Normal Fuel Delivery** | too much fuel

Fuel pressure has a direct impact on fuel delivery volume at a given injector pulse width.

- Insufficient fuel pressure and volume can be caused by low fuel pump voltage, a worn pump, a leaking hose between pump and sender unit inside the tank, a clogged fuel filter, or a faulty fuel pressure regulator. Check fuel pump current draw by using a low-amp current probe attached to a digital voltmeter or oscilloscope to verify (or predict) fuel pump failure.

Wait! Won't the oxygen sensor make corrections to maintain lambda 1.0? That depends. If the system goes too far rich or lean, the feedback/fuel control system may not be able to correct the mixture.

## Fuel Supply Testing

There are several ways to test the fuel delivery system.

- Check fuel pressure
- Check fuel pump volume
- Check fuel pump current

The procedures used for these tests are a book in themselves. They are also potentially dangerous at times, since some require disconnecting fuel lines to connect fuel gauges or flow meters. Fuel trapped inside the lines may be highly pressurized.

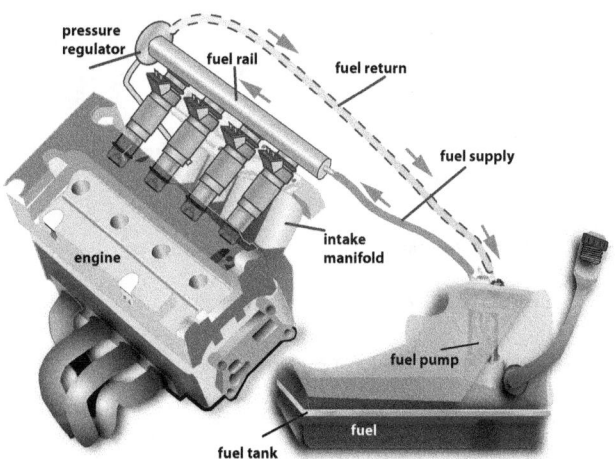

Fuel supplies have changed dramatically over the last 20 years. Most fuel pumps are now located inside the fuel tank. Pressure regulators may be at the rail, or inside the fuel tank; others hide inside the fuel filter. The Fuel Return Line, once a standard feature, is fast disappearing as returnless, on-demand fuel systems gain favor. Fuel pressures are way up, especially in engines using Gas Direct Injection (GDI). Be safe! Use only the appropriate equipment and approved test procedures to avoid vehicle damage or personal injury.

DTCs commonly associated with excessive fuel trim and mixture control failures:

**P0171, P0174** - Mixture Lean

**P0172, P0175** - Mixture Rich

**P0191 - P0194** - Fuel Rail Pressure Sensor Faults

# Troubleshooting Tips

## OBD II Vehicle Repair Sheet - Pre-Repair Evaluation

In the hospital, all patients have a chart that includes a pre-treatment list of vital signs. A chart like this—or one of your own design— is a good way to track diagnosis and repair, and verify a good result.

Date_____Customer Name_____

Phone_____

Vehicle Type_____Model_____

Engine_____Mileage_____

VIN (Vehicle Identification Number) __ __ __ __ __ __ __ __ __ __ __ __ __ __ __ __ __

Current MIL status:

MIL ON with no driveability complaint_____

MIL-ON accompanied by driveability complaint (List symptoms)

_____

_____

_____

DTCs Stored:

DTC  1)_____Descriptor_____

DTC  2)_____Descriptor_____

DTC  3)_____Descriptor_____

DTC  4)_____Descriptor_____

DTC  5)_____Descriptor_____

# Troubleshooting Tips

**Freeze Frame Data (Fill in all spaces before erasing DTCs; or attach printout)**

Fuel System Status    Open Loop _____ Closed Loop_____

Calculated Load___%

Short Term Fuel Trim Bank 1_____%              Bank 2_____%

Long Term Fuel Trim Bank 1_____%    Bank 2_____%

Engine Speed_____RPM

Vehicle Speed_____MPH

Engine Coolant Temperature_____degrees (F/C)

If single-cylinder misfire is detected, indicate cylinder(s)_____

**Current Parameter Display Information (compare to Freeze Frame)**

Fuel System Status    Open Loop_____ Closed Loop _____

Calculated Load_____%

Short Term Fuel Trim Bank 1____%    Bank 2_____%

Long Term Fuel Trim Bank 1_____%              Bank 2_____%

Engine Speed_____RPM

Vehicle Speed_____MPH

Calculated Load_____%

Engine Coolant Temperature_____degrees (F/C)

Visual and Maintenance Inspection (Check fluid levels and note any abnormal conditions)

1) Engine oil level and condition_____

2) Battery Voltage_____Charging Voltage_____

3) Condition of all belts and hoses_____

4) Note any and all additional maintenance and repair items that indicate a potential problem:

# Troubleshooting Tips

## A Step-by Step Repair Sequence

Here's a quick repair sequence that solves most MIL-on conditions.

### 1) Retrieve and Record DTCs
Look up the DTC definition, if possible. This is EXTREMELY important. You can't fix a problem unless you know exactly what caused it. Looking up the exact DTC definition is time well spent. Skipping over this part of the diagnosis can cost you time in the long run. (See pages 46-50.)

### 2) Look at Freeze Frame
Do NOT erase DTCs until you view and record Freeze Frame data. CSI teams never throw away forensic evidence. Erase codes and you erase all the evidence from the crime scene. (See page 52.)

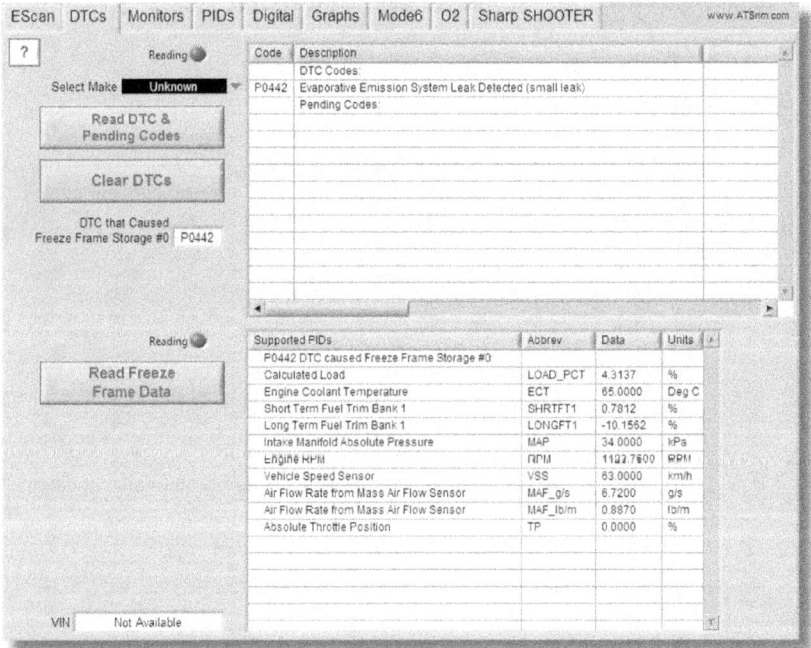

Read and record DTC and Freeze Frame data. One of the nicest features of a PC-based scan is that you can save screen images like this one quickly.

# Troubleshooting Tips

## A Step-by Step Repair Sequence

### 3) Carefully Review Datastream

Scrutinize PIDs and values. Look for anything that seems out of place or illogical. Compare data. (See Chapter 5: **Datastream.**)

Ask these questions:

- Do MAP and BARO agree KOEO (Key-On Engine-Off)? Does the BARO data PID match the actual barometric pressure at your location? Look for a YES answer to both.
- Are IAT and ECT the same in a stone cold engine KOEO? Do they match the ambient temperature (the temperature of the surrounding air)? Look for a YES answer to both.
- Is battery voltage good KOEO?
- Is system (charging) voltage good KOER?
- Does the fuel system enter closed loop quickly?
- Do changes in short term fuel trim indicate that the computer is in control of fuel delivery?

You're the doctor. Review the patient's vital signs carefully.

| Supported PIDs | Abbrev | Data | Units |
|---|---|---|---|
| ✓ Calculated Load | LOAD_PCT | 21.1765 | % |
| ✓ Engine Coolant Temperature | ECT | 201.2000 | Deg F |
| ✓ Short Term Fuel Trim Bank 1 | SHRTFT1 | 5.4687 | % |
| ✓ Long Term Fuel Trim Bank 1 | LONGFT1 | 3.9062 | % |
| ✓ Intake Manifold Absolute Pressure | MAP | 7.6778 | HG |
| ✓ Engine RPM | RPM | 691.0000 | RPM |
| ✓ Vehicle Speed Sensor | VSS | 0.0000 | mph |
| ✓ Ignition Timing Advance for #1 Cylinder | SPARKADV | 4.5000 | deg |
| ✓ Intake Air Temperature | IAT | 129.2000 | Deg F |

Don't forget to compare KOEO (key-on, engine-off) and KOER (key-on, engine-running) data. Too many diagnosticians fail to capture a screen of data **before** the engine is started. Side-by-side data comparisons of KOEO and KOER data effectively highlight many problems.

Think about it: most PID values should change with the engine running. Some, like rpm, will change quickly. Others, like ECT, change gradually as the engine warms. Learning to recognize "normal" data is the first step in identifying "abnormal" data associated with a problem.

The Auto Emissions Bible

# Troubleshooting Tips

## A Step-by Step Repair Sequence

### 4) Baseline the Vehicle

Baseline the vehicle. This has already been mentioned repeatedly, with good reason. We are always amazed to find an "unfixable" car put right with a fresh fuel filter or a set of spark plugs. Or a car with multiple, random DTCs that has a loose ground connection. Pick low hanging fruit first. Perform a thorough visual inspection of all belts, hoses, accessible filters, vacuum hoses, air intake ducts and hoses, and fluid levels. (See page 159.)

If datastream points to a specific problem, dig deeper using standard shop repair equipment and common system tests.

Do you have a P0300 DTC (random misfire), with Freeze Frame that indicates misfire under load? Mist the secondary ignition wires using an old window cleaner bottle filled with water containing an ionizer like baking soda. This will help you locate secondary insulation that leaks voltage to ground. No need for rocket science when simple tests will do.

This Ford Contour timing belt jumped time after losing more teeth than an NHL goalie. A simple engine vacuum gauge pointed us in the right direction. Even though the engine still ran, the engine vacuum at idle was an anemic 9 inHg. The engine did not misfire, but we could outrun the car on foot; we started looking for a timing issue.
(See pages 174-175.)

### 5) Fix the Basics First

We won't get all preachy about this, but dagnabit, if you see a battery terminal that looks like a Chia Pet®, clean the darn thing. First! If the engine has 110,000 miles on the factory spark plugs and a P0300 (random misfire code) then you need to look at the _____ (fill in the blank). (See pages 160-163.)

The Auto Emissions Bible

# Troubleshooting Tips

## A Step-by Step Repair Sequence

### 6) Do Your Research

Auto repair is not for dummies. It is time to put that misconception to bed, permanently. The successful modern repair professional is comfortable with a personal computer and the internet, reads for comprehension, and knows how to use math and science. Most importantly, he has superior analytical skills.

The repair professional also knows that information is now an essential repair tool. Good techs are never ashamed to look in a book, long considered by some as a sign of weakness! Read the book with pride! Auto repair is not for dummies, but guessing is!

Access data from:

- **Repair databases like Mitchell and AllData.** These repositories of repair information provide easy access to vehicle specific specifications, test standards, and sequential how-to procedures. Both companies offer extended service contracts for repair professionals, or reasonable day rate charges for prosumers.

- **Tech support hotlines.** Identifix specializes in both live tech support or 24/7 access to internet reference materials that include case studies, pattern failure data, and a reference library. Sometimes, there is no substitute for a friendly, knowledgeable voice at the other end of the phone line.

- **The National Automotive Task Force is a not-for-profit organization that provides an online portal to vehicle maker (OEM) web sites.** OEM sites sell reference materials and vehicle specific repair manuals. They are also the storehouse for software upgrades required when PCM reprogramming is the only correct repair procedure. (See pages 278-279.)

The Auto Emissions Bible

193

# Troubleshooting Tips

## A Step-by Step Repair Sequence

### 7) Check TSBs

Technical Service Bulletins (TSBs) have moved up the food chain several rungs since OBD II was introduced. The TSB is a white paper informing dealerships of pattern failures, approved and revised repair procedures, and updated parts and information.

Checking TSBs related to vehicle symptoms should be part of any good repair procedure. There are many reasons to do this, including increased vehicle complexity and constant revisions to OEM software.

**Important: There are vehicles that cannot be properly repaired without a PCM reflash.** There are vehicles on the road today that will extinguish the dreaded MIL as soon as a new software package is installed in the PCM. Failure to identify this problem—and solution—early in your diagnosis will lead to hours of wasted time and frustration.

This trend will not diminish. Electricity is as important to vehicle operation as fuel inside the tank. Witness the addition of by-wire changes to throttle control and power steering. Engineers increasingly rely on calibration changes to correct design problems, or to compensate for performance issues caused by normal wear and tear.

Reprogramming requires a special J2534 hardware interface, a computer that meets a set of minimum standards defined by the vehicle maker, and the correct software revision. Software revisions are available for a fee from each vehicle maker. (See page 276.)

# Troubleshooting Tips

## A Step-by Step Repair Sequence

### 8) Check Vehicle History

Vehicle history can be a useful source of clues when you solve a repair mystery. Recently, we saw a vehicle with an overall performance problem that had defied multiple previous repair attempts. Our iVac (pages 176-177) showed low but steady engine vacuum. Repair history indicated that the timing belt had been replaced right before the problems started—and then repeatedly checked for proper installation by multiple parties.

Assured by everyone involved that the timing belt was not the issue, we immediately removed the timing cover and discovered that the cam timing marks **were improperly aligned**. When you've tested the basics repeatedly, it often pays to check them again.

### 9) Fix the Car

By now, you should have a plan. Make your repairs. Then compare post-repair test results to data gathered in steps 1-4. You should see a difference in pre- and post repair data as a verification that your repairs are successful.

Those of you wanting to use Mode 6 for repair verification should jump to **Chapter 12: Mode 6.**

In the next chapter, we look at ways to make sure your vehicle passes an emissions test after the repair is completed. We'll also discuss what to do after a vehicle fails an emissions test.

## Troubleshooting Tips Review

- All emissions repairs should reference the VIR, and baseline the vehicle to identify its condition. Baselining provides a benchmark for post-repair comparison and repair confirmation.

- Check and eliminate basic problems first. Include basic electrical system tests in all diagnoses.

- Identify the fuel system: MAP or MAF. Match test results to system characteristics.

- Purchase and use a vacuum gauge regularly. It is inexpensive and extremely effective for locating clogged exhausts, misfire, vacuum leaks, and engine mechanical problems.

- Oxygen sensors are used by the PCM for fuel control and tests of multiple vehicle subsystems.

- Fuel trim is a free and highly effective diagnostic tool built into the generic OBD II diagnostic suite. Use it.

- Write and use a vehicle worksheet and establish a regimented routine. Organized work flows are highly effective, add discipline to your diagnostic procedures, and eliminate missed steps that leave gaping holes in diagnoses.

# PASSING THE EMISSIONS TEST

# 11

# Passing the Emissions Test

## Post Repair Vehicle Prep

Making sure the car is ready for the scan tool emissions test is an ongoing challenge for vehicle owner and repair technician alike. Even if the MIL is off, a vehicle will still be turned away from the emission test center for incomplete monitors.

On the surface, monitor readiness is not a complicated issue: When non-continuous monitors run all the way, they change from incomplete to complete (Not Ready to Ready). If monitors are incomplete, the vehicle cannot be tested until they are.

Here are a few things worth considering as you prepare a vehicle for a scan tool emissions test:

• **Think twice before erasing DTCs.** Many of us go on "auto pilot" and routinely erase DTCs following any MIL-related vehicle repair. Granted, erasing DTCs may be the right thing to do. But pause, take a deep breath, and think it over: erasing some DTCs may be the worst thing we can do, since it resets *all* monitors to incomplete (Not Ready) status.

If you are sure you fixed the problem, and it is a simple one like a loose gas cap, why not let the PCM turn off the MIL? It may take a lot less driving to turn off the MIL than it does to run a catalytic converter monitor in a car with 170,00 miles on it.

> Remember, the PCM runs the monitors and will turn off the MIL if the test that stored the DTC runs and passes on three consecutive trips. The MIL will go off, already completed monitors will still be displayed as Complete (Ready) and the vehicle will pass the emissions test.
>
> **CAUTION:** This works only if the monitor that set the DTC will run with DTCs in memory. Some will, some won't.

• Make a plan for the vehicle. Assuming the vehicle is allowed one or even two incomplete monitors, and the monitors have been reset to incomplete (Not Ready) by erasing DTCs, it may **not** be necessary to run all of the monitors. Run enough monitors to satisfy the requirement for complete monitors, and get the vehicle tested (assuming the MIL is not on).

The Auto Emissions Bible

# Passing the Emissions Test

## Post Repair Vehicle Prep

- **If you are confident that you have repaired the vehicle fault that stored the DTC, see if the vehicle owner is willing to drive the car to run the monitors.** This option works best when there is no imminent deadline for passing the test and renewing the registration, and when you have a good relationship with a customer who is willing to work with you.

  If it's your car, no problem!

- **Somebody has to drive the vehicle in ways that let monitors run.** For example, let's say that someone has replaced a dead battery in a vehicle and the PCM "blackout" reset all the monitors to incomplete. And let's also say that this car is used exclusively for milk and eggs grocery store runs at speeds of 35 mph, or less. If monitor enabling criteria for things like EGR and catalyst tests require steady throttle cruise at speeds of 55-65 mph, those monitors will NEVER run. The vehicle will certainly be turned away at the emission test center for incomplete monitor status.

  If you are a repair professional, you may need to charge the vehicle owner for your time and drive the car so enabling criteria are met and monitors can run. In some cases, you can accomplish all needed driving in the safety and convenience of your own shop.

- **Once a non-continuous monitor runs to completion at least one time, it will be displayed as Complete (Ready)** until someone erases DTCs with a scan tool, or the PCM Keep Alive Memory (KAM) is erased with a scan tool (or by removing battery power to the PCM).

> This may sound simplistic, but given the opportunity, take the emissions test when there is nothing wrong with your car. Here in Ohio, we test vehicles every other year. My 2000 model is tested in even years because it is an even year model. It is currently January of an even year, and there is currently nothing wrong with the car. It will pass easily. Test it now, and it won't have to be retested for two more years.
>
> By electing to test a car with no problems as soon as possible in a test cycle, you avoid last minute desperation repair scenarios that always seem to cost more in cash, aggravation, and inconvenience.

# Passing the Emissions Test

## Running Monitors

Here is a common dilemma: Somebody erased all emissions data and reset the monitors, and you need to complete the monitors before taking the vehicle to the test center for a scan tool emissions test. Let's look at our options.

### The Drive Cycle

Trips, monitors, and drive cycles are vehicle- and monitor-specific. We will loosely define a **Drive Cycle** as a *set of driving conditions that let a monitor run to completion*. Some folks also refer to a "global" drive cycle that combines conditions that let **all** vehicle monitors run to completion.

There is no "one-size-fits-all" drive cycle that will run all monitors on all vehicles. Each monitor requires a set of enabling criteria before it will run. Keep it simple; *enabling criteria are conditions needed to run a monitor*. Many monitor enabling conditions are similar enough to let us set up a Generic Drive Cycle that works on many vehicles.

Here is a **Generic Drive Cycle** that completes the non-continuous monitors in most vehicles, in about 30 minutes. (This has been tested in several states on real cars, and feedback has been positive.)

**Step 1.** Start with a cold vehicle; ideally, let it sit for 6-8 hours before the test, without a start. (This is primarily for EVAP monitors in some makes that require an extended cold soak of up to 6 hours as part of the EVAP monitor enabling criteria. Especially some Fords.)

**Step 2.** Connect a scan tool.

**Step 3.** Start the engine and warm it to normal operating temperature.

**Step 4.** Drive the vehicle for 10 minutes at highway speeds. Take a passenger along to watch the monitor status display and inform you as monitors run to completion

**Step 5.** Drive the vehicle for 20 minutes in stop and go traffic with at least four idle periods. **Note:** A few monitors on some makes run only after the engine is shut down. This is more common now that we have engine off fuel system leak detection monitors. Other monitors will run but not update their status to complete until the ignition is switched off, and then switched on again.

# Passing the Emissions Test

## Drive Cycle Special Notes

If the generic drive cycle doesn't work, it's research time. Look up the OEM-specific drive cycle for the vehicle and follow it; to the letter! There are a few quirky drive cycle patterns out there, so be warned.

Prosumers: See if your local library has library cardholder access to a repair database. It could be a lifesaver.

Over the years, some diagnosticians have recommended erasing PCM memory by disconnecting battery power. The theory is that some monitors run in a faster mode when all learned data are erased. This may be true in some cases.

Be careful with this one.

**Here's why:** Disconnecting the battery in some vehicles erases all "drive by wire" learned data. The vehicle may refuse to idle until a scan tool with special software "relearns" the computer. And don't forget all the other stuff like radio and seat position presets that commonly disappear from volatile memory when power is removed from the PCM's keep alive memory circuits.

# Passing the Emissions Test

## Using an OEM Drive Cycle

The example below is a drive cycle specifically designed to run the EGR monitor in several vehicles from the same vehicle maker. It is similar in many respects to the generic drive cycle.

This OEM refers to enabling criteria as "preconditions." Note that one of the entry conditions listed for this monitor is that the MIL must be off. To eliminate other conditions that might result in false test results, the monitor will not run at extremely high altitude or at very low ambient temperatures. The monitor is also suspended when ECT is cold.

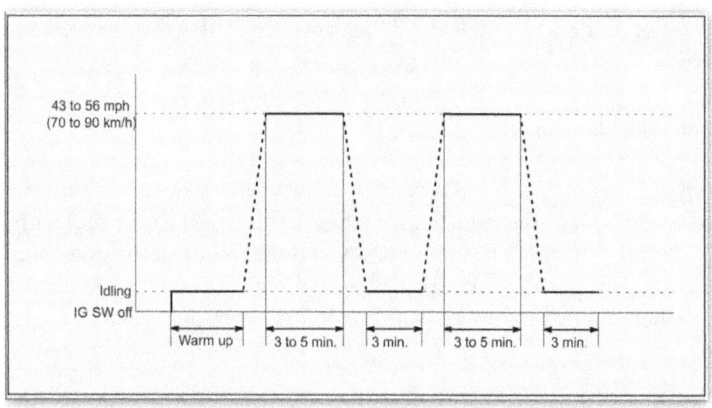

**Preconditions (Enabling Criteria)**
This monitor will not run unless:
   • The MIL is OFF.
   • The test is run at altitudes below 7870 ft.
   • The intake air temperature (IAT) is 14°F (-10°C) or higher.
   • The engine coolant temperature sensor (ECT) is 167 °F (75 °C) or higher.

**Drive Cycle**
   (1) Connect a scan tool.
   (2) If IAT is less than 50°F (10°C), idle the engine for 10 minutes.
   (3) Drive vehicle at 43-56 mph for a period of 3-5 minutes.

**Special Instructions**
   (4) Stop vehicle and let the engine idle for 3-5 minutes.
   (5) Repeat steps 3 and 4.

If monitor does not change to Complete, repeat steps 3-5.

Let's look at another example on the next page.

The Auto Emissions Bible

# Passing the Emissions Test

## Practice Exercise

A vehicle has failed an emissions test for an illuminated MIL.

- All installed monitors are listed as Complete in the Readiness List.
- A DTC of P0112 (IAT Low) is stored. It is the only DTC.
- The vehicle owner has received a 10-day extension to perform repairs.
- Freeze Frame and current datastream both show an IAT value of 135°C (275°F).
- Further tests indicate that the IAT is shorted.
- A new IAT is installed.

**Question:** What is the first place to look to determine if the new IAT fixed the problem?
**Answer:** Datastream, to see if the IAT/circuit is performing correctly.

**Question:** Should we erase DTCs?
**Answer:** Use the chart below to make your decision.

| Fault | DTC | DTC Definition | Test Values |
|---|---|---|---|
| IAT Low Input | P0112 | Continuous short to ground in IAT or IAT circuit | IAT=135°C (275°F) |
| | | | |
| **Enable Conditions** | **Test Time/Frequency** | **MIL Test Type (A or B)** | Special Notes |
| No ECT DTCs No VSS DTCs VSS>25 mph Engine run time 10 sec | Continuous 175 failures in 1200 samples | Continuous Type B | > = greater than < = less than |

We could erase DTCs. But that might not be the best move. Instead we will give the PCM a chance to test the IAT. If the monitor passes on three trips, the PCM will turn the MIL off.

Assuming the MIL goes out, the DTC will remain in memory for a while—but that will not result in a test failure as long as the MIL is off. If we do not reset the monitors, we will not need to run them again. This saves time.

**Question:** Will the monitor run if we start the engine three separate times without driving the vehicle?
**Answer:** No. The monitor for this vehicle and emissions package requires a vehicle speed sensor (VSS) input greater than 25 mph as part of a trip.

# Passing the Emissions Test

## Monitor Exercise

This chart shows a sample set of test limits and enabling criteria. Use the information on this page to answer the questions on the next page.

| PID/Fault | DTC | DTC Definition | Test Values |
|---|---|---|---|
| EVAP Gross Leak | P0440 | Test detects a missing gas cap or a gross leak in the EVAP system | EVAP leak > 0.040" |

| Enable Conditions | Test Time/Frequency | MIL Test Type (A or B) | Special Notes |
|---|---|---|---|
| No MAP DTC<br>No MAF DTC<br>No TP DTC<br>No IAT DTC<br>No O2 DTC<br>No VSS DTC<br>No Misfire DTC<br>No Fuel Trim DTC<br>No Injector DTC<br>No EGR DTC<br>No ECT DTC<br>No AIR DTC<br>BARO>75.20kPa (8000ft)<br>$4°C \leq$ Powerup ECT $\leq 30°C$<br>$4°C \leq$ Powerup IAT $\leq 30°C$<br>$\Delta$ECT-IAT $\leq 8°C$<br>$\Delta$IAT-ECT $\leq 8°C$<br>15% <Fuel Level < 85%<br>5.0V < System Voltage < 18V | Test begins 180 seconds after start and ends when tank vacuum reaches -7.9 $inH_2O$ or timer expires (37 seconds) | Type B | > = greater than<br>< = less than<br>$\leq = \lambda$ less than<br>   or equal to<br>$\Delta$ = Indicates difference<br>between two values |

## Legend

$4°C = 39.2°F$

$\Delta8°C = 14.4°F$ ($\Delta$ indicates the difference in temperature)

$30°C = 86°F$

$75.2kPa = 22.20$ inHg

Inches of water is a measurement scale used to measure low pressures. One psi is equal to 27.68 inches of water (inH2O).

Type B indicates that this code does not turn on the MIL in a single trip.

The Auto Emissions Bible

# Passing the Emissions Test

## Monitor Exercise

**Question:** When does this monitor run?
**Answer:** 180 seconds after engine start up, assuming all other conditions are met.

**Question:** What size leak is the monitor testing for?
**Answer:** 0.040 inch

**Question:** When does the test pass?
**Answer:** When EVAP system pressure sensor indicates 7.9 $inH_2O$ of vacuum.

**Question:** Will this monitor run with the engine at normal operating temperature?
**Answer:** NO. ECT must be below 86°F for the monitor to run.

**Question:** Will this monitor run if the tank was just filled?
**Answer:** No. The monitor is suspended if the fuel tank level is below 15 percent or above 85 percent.

**Question:** Will this monitor run above 8000 feet altitude?
**Answer:** No. The monitor runs only at altitudes below 8000 ft.

**Question:** Will this monitor run if IAT is 32°F?
**Answer:** No. The ECT must indicate a temperature between 39.2 and 86°F.

**Question:** Will this monitor run if ECT is 130°F and IAT is 50°F?
**Answer:** No. ECT must be between 39.2 and 86°F; ECT and IAT must be within 14.4° F of one another.

**Question:** The monitor stops when it detects -7.9 in $H_2O$. What does this reading indicate?
**Answer:** That the EVAP system is sealed well enough to pull the minimal test vacuum, indicating that there is no gross leak.

**Question:** Assuming there are no other problems with the EVAP system, why might the PCM fail to see -7.9 in $H_2O$ during the test?
**Answer:** Fuel vapor pressure generated inside the tank reduces vacuum even if there is no leak. Test parameters are very tight on this test to eliminate the chance that special conditions like high ambient temperature or fuel slosh will create excessive amounts of system pressure.

**Question:** How do we interpret the line "BARO > 75.20kPa (8000 ft)"?
**Answer:** Barometric pressure decreases at higher elevations, so the PCM looks for a *higher* barometric pressure reading as a sign that the vehicle is *below* an elevation of 8000 ft.

# Passing the Emissions Test

## Common Emission Test Failures

Emission test data collected since OBD emissions tests started have shown that vehicle test failures fall into some fairly predictable categories. And many of those failures are traced not to some exotic computer glitch, but to a lack of maintenance that commonly includes low-tech problems like worn spark plugs, leaking plug wire insulation, neglected cooling systems, and clogged filters.

We apologize for being trite, but: If you hear hoofbeats, think horses, not zebras. Look for simple answers to vehicle problems first. Leave the rocket science on the back burner until all the common reasons for a DTC have been thoroughly examined and eliminated.

Let's look at one test program and its accumulated data on common test failures. The high probability codes and system failures shown here are representative of those in other states.

> If you hear hoofbeats, think horses, not zebras.

# Passing the Emissions Test

## Oregon Top Ten DTCs

Gary Beyer, engineer at the Oregon DEQ, was kind enough to share data collected from their vehicle emission program. It includes a Top Ten list of common DTCs.

The results are similar to those found across the country, and most fall into predictable categories, many of which can be diagnosed with generic test equipment found in professional repair facilities and many prosumer garages across the country. The numbers below are from 2009 Oregon vehicle test data.

Problems are listed from most to least common, including actual failure numbers, shown in parentheses. Each is explained in greater detail on the next few pages.

---

### Oregon's Top Ten DTCs

**1)** System Too Lean (4226)

**2)** Catalyst System Efficiency Below Threshold (4181)

**3)** Knock Sensor 1 Circuit (2836)

**4)** $O_2$ Sensor Heater Circuit (2447)

**5)** Exhaust Gas Recirculation Flow Insufficient Detected (2357)

**6)** Random/Multiple Cylinder Misfire Detected (1888)

**7)** Evaporative Emission System Leak Detected - large leak (1664)

**8)** Evaporative Emission System Leak Detected - small leak (1638)

**9)** $O_2$ Sensor Circuit Slow Response (1548)

**10)** $O_2$ Sensor Circuit No Activity Detected - (1145)

---

# Passing the Emissions Test

## Oregon Top Ten Emission Failures

**1) System Too Lean** - A perennial favorite, **system lean** DTCs indicate that the balance between air and fuel is incorrect: the ratio has too much air or too little fuel. Common causes include a faulty MAF, unmetered air leaks downstream of the MAF, engine vacuum leaks, clogged fuel filters, restricted fuel injectors, weak fuel pumps, or a fuel regulator that does not maintain sufficient fuel pressure.

**2) Catalyst System Efficiency Too Low** - The vehicle computer has determined that the catalytic converter efficiency has fallen below an acceptable level. These monitors are usually accurate, assuming there are no other vehicle problems that could skew the test results, such as a sleepy oxygen sensor. The algorithms used to test the catalyst are patient and cautious in their testing. No vehicle maker wants the MIL on for a failed catalyst, so these monitors will repeat tests using special algorithms to avoid false MIL illumination. See pages 182-183.

> **Catalyst Quick Notes:** Check for Technical Service Bulletins before condemning a catalyst. Catalyst "failures" sometimes result from software programming errors, and some software upgrades loosen test standards, allowing the catalysts to pass, once the PCM is reflashed.
>
> Just as important: Always find out why the catalyst died in the first place. Misfire and too much fuel inside the cylinders both kill catalysts fast. Catalyst poisoning by oil and antifreeze will coat and reduce the effectiveness of the catalyst's active materials; catalyst killers must be found, and stopped.

**3) Knock Sensor** - Knock sensors detect detonation inside cylinders caused by insufficient gasoline octane or excessive cylinder pressure. Assuming you don't have that 400 watt sound system set to "stun," ping or knock may be audible inside the car, especially during part throttle acceleration up a grade. Its sound is similar to the noise made by a chain drive as it drags a roller coaster car up the lift hill to the first peak. The noise indicates that fuel is exploding violently inside the cylinders, a condition that can cause severe engine damage, as well as harmful emissions.

The knock sensor detects vibration caused by the explosions and sends a signal to the computer, which then retards ignition timing in steps until the knock or ping ceases. Considering the potential for expensive engine damage, this is one code condition that should be investigated and cured quickly.

## Oregon Top Ten Emission Failures

**4) $O_2$ Sensor Heater Circuit** - OBD II oxygen sensors have electrical heaters that get them toasty hot and ready to run in seconds. Older sensors with no internal heater were heated only by exhaust gases, and took their sweet time coming on line. Without heaters, some sensors may never reach operating temperature due to their location in a cooler section of the exhaust, farther from the engine. Don't replace the sensor until you test the heater, its wiring, and circuit connectors for an open circuit. Some heaters are fused, so check that, too.

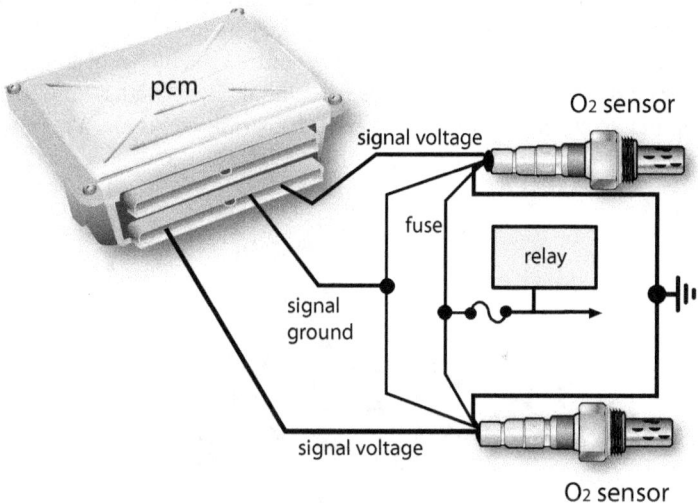

**5) Exhaust Gas Recirculation Insufficient Flow** - EGR systems characteristically clog over time with heavy black deposits that can jam up the EGR valve or restrict exhaust flow through the EGR piping and EGR ports inside the cylinder heads. This is a common and long standing problem. An equally low tech cure is to remove the EGR valve and connecting tubes, dig out the hard deposits with a pick and a bottle brush, and flush what's left with a liquid decarbonizer.

**CAUTION:** Easy on the liquids. We don't want to bend connecting rods by cranking over an engine with liquid inside the cylinders. If there is any chance that your EGR flushing has left liquid cleaner inside the cylinders, play it safe: pull the plugs and turn the engine over by hand, two full turns, to push any liquid cleaner from the cylinders.

# Passing the Emissions Test

## Oregon Top Ten Emission Failures

**6) Random/Multiple Cylinder Misfire Detected** - The all too common P0300 DTC identifies a cylinder misfire condition, but does not identify the exact cylinder(s) by number. This generally means a couple of things: (1) the misfire really is a general condition affecting multiple cylinders, or (2) the OBD system does not have the software coding to identify cylinder-specific misfire. Some OBD systems can do this; others cannot.

The common causes list for P0300 is a long one. Include: Worn, shorted, or fouled spark plugs, leaking secondary ignition insulation, defective ignition coil(s), low coil voltage, leaking fuel injectors, clogged fuel injector(s), contaminated or poor quality fuel, a massive vacuum leak, blown head gasket(s), or other engine mechanical problems resulting in low cylinder compression. Remember, start simple. If the plugs have 100k miles on them, a new set is an inexpensive and logical starting point.

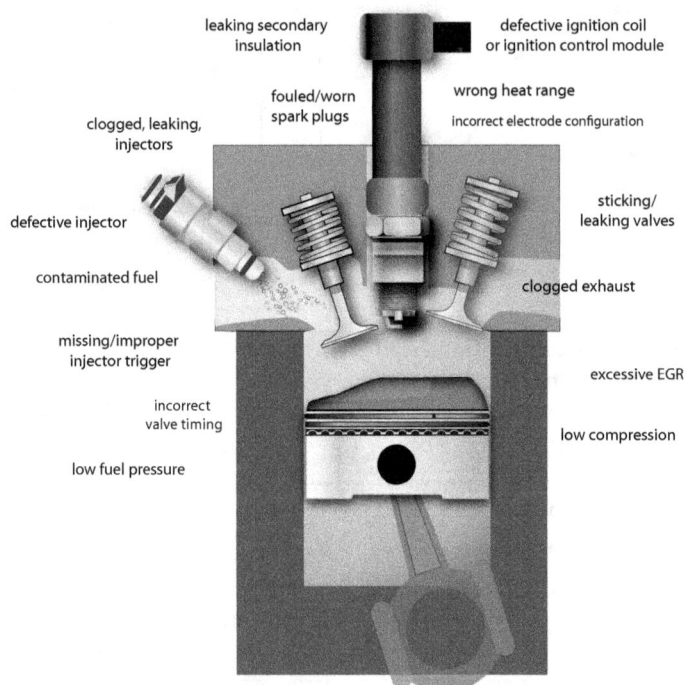

Loose accessory belts and/or weak belt tensioners can cause misfire DTCs.

## Oregon Top Ten Emission Failures

**7) and 8) Evaporative Emission System leaks, both large and small** - We'll group these together, since the root cause for these EVAP DTCs is a leak (or leaks) in the fuel containment system that lets raw fuel vapor escape to the atmosphere. If the code number indicates a large leak, make sure the gas cap is on, and tight. Happens all the time. If there is a small leak, things get tougher, because you could be looking for a teeny, tiny hole 0.020 inch in size.

Diagnostic procedures commonly include applying an inert gas like nitrogen or carbon dioxide to the system through a test port or fill neck adapter, and then looking for bubbles as you apply soapy water to all fuel and fuel vapor hoses, connections, etc. Smoke machines are popular with professional repair technicians, since they pump thick smoke into the fuel system that is easier to spot as it escapes from a leak. Finding small leaks can be a test of your patience, but it isn't normally a high-tech process. (See pages 122-125.)

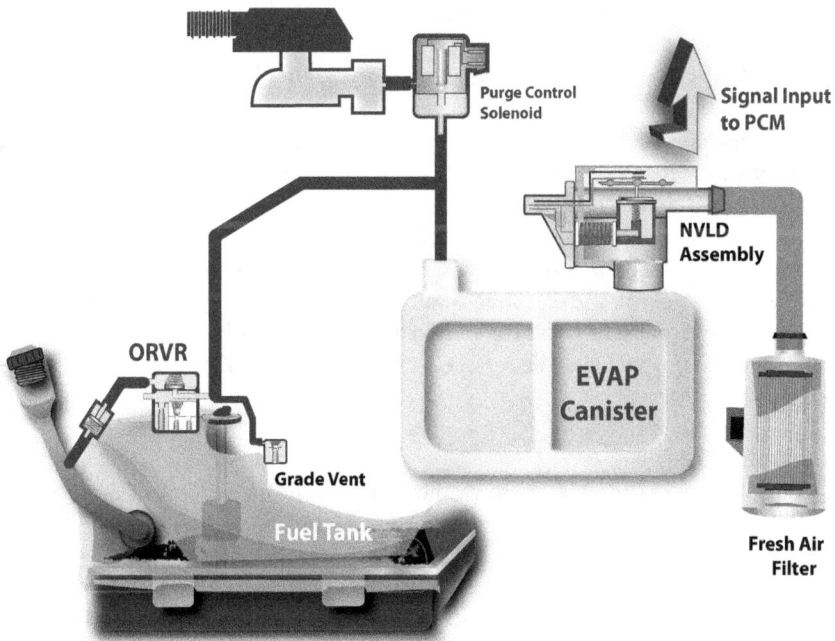

Purge Control Solenoid

Signal Input to PCM

NVLD Assembly

ORVR

EVAP Canister

Grade Vent

Fuel Tank

Fresh Air Filter

# Passing the Emissions Test

## Oregon Top Ten Emission Failures

**EVAP Test Cautions:**
We do not recommend applying compressed air to a closed fuel system for several reasons:

1) Pressurized air + fuel vapor escaping at high velocity from a leak + a spark = BOOM. This is just too risky to try. It's why we recommend an inert gas.
2) Too much pressure can damage fuel system seals and valves. Please apply only as much pressure as the vehicle maker recommends for test purposes, normally 1 psi, maximum.
3) Compressed air commonly contains a lot of moisture and even dirt.

**9) $O_2$ Sensor Slow Response** - Oxygen sensors are voltage generators. As they age, they may continue to generate the correct voltages, but respond more slowly to changes in exhaust oxygen. Accurate tests involve exact measurement of voltage changes, measured with an oscilloscope, a procedure outside the scope of this book. Most of the time, a new sensor is the cure, especially on high mileage cars. Oxygen sensors that die too soon, however, often die early due to contamination from lube oil or coolant in the exhaust, each indicating a more serious engine mechanical problem.

**10) $O_2$ Sensor No Activity Detected** - Look for an open wire, shorted wire, or dead sensor. A good shortcut is to look at the sensor on a scan tool with the ignition key turned to the on position. If the sensor voltage in datastream is nailed at a steady zero volts, the sensor or signal wire between the sensor and PCM is shorted to ground. Disconnect the wire harness at the sensor and check the datastream value again. If sensor signal voltage shoots up to about half a volt (or bias voltage), the sensor itself is short-circuit to ground and must be replaced. (See pages 178-183 for more on oxygen sensor tests.)

### Summing Up

A few of these issues are scary to the home mechanic or vehicle owner because components like catalytic converters are expensive. But many DTCs can be eliminated with recommended maintenance and simple repairs using common tools and equipment found in any reputable repair shop and in many home garages.

Next, lets look at which **systems** are responsible for the most DTCs.

# Passing the Emissions Test

## Oregon Top Ten Emission Failures

### DTCs Grouped by Subsystem

Next, let's look at the Oregon test numbers again, but this time we'll group failures by *subsystem*.

**P04xx Codes - Emission Controls** come in first, accounting for 34% of all DTCs. Common components in this group include: Exhaust gas recirculation (EGR), evaporative emission components (EVAP), catalytic converters, secondary air injection (AIR), and cooling fan speed controls.

**P01xx and P02xx - Fuel and Air Metering** comes in a close second, accounting for an additional 33% of all DTCs. This list includes all components used to measure air density, air and engine temperatures, fuel pressure and injector control circuits, oxygen sensors and their heaters, and forced air components like turbo- and superchargers.

**P03xx - Misfire Codes come in third**, accounting for 26% of all DTCs. (If you're keeping track, simple addition should show that three vehicle subsystems account for 93% of all vehicle DTCs recorded by the Oregon vehicle test program.) P03xx codes identify actual misfire, both random and cylinder-specific, component problems in ignition coils, and timing components like crankshaft and camshaft sensors.

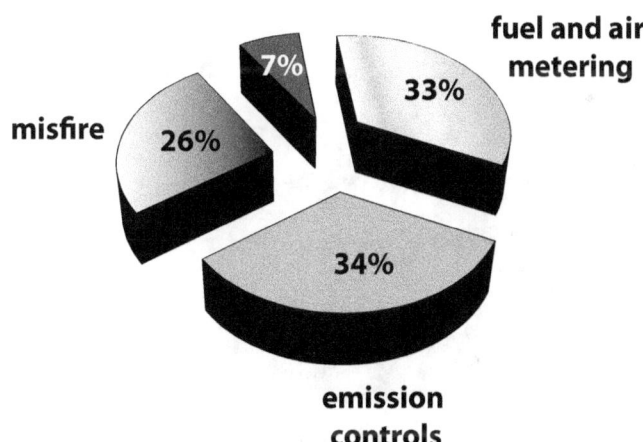

# Passing the Emissions Test

## Oregon Top Ten Emission Failures

### And the Rest?

Bringing up the rear, we have three more subsystems: transmissions, vehicle speed inputs, and computers: lumped together, they account for a measly 7% of all remaining DTCs. Computers and their outputs account for only one percent of all DTCs!

Let's make a few general assumptions based on these numbers:

- **Motorists can take some solace in knowing that many problems occur in vehicle subsystems that respond well to low-tech replacement of common, less costly service items like spark plugs and filters.** Many emission system DTCs are eliminated by tightening a gas cap, or washing away carbon deposits from EGR passages.

- **Any good news for repair shops? Sure. Any reputable repair shop with professional test equipment and an ongoing commitment to training, can repair a high percentage of fuel, misfire, and emission component failures.** That smaller block of transmission and internal computer repairs is already jobbed out to repair specialists who have niche-market training and equipment.

Fears that OBD repairs are the exclusive domain of rocket scientists may be somewhat overblown.

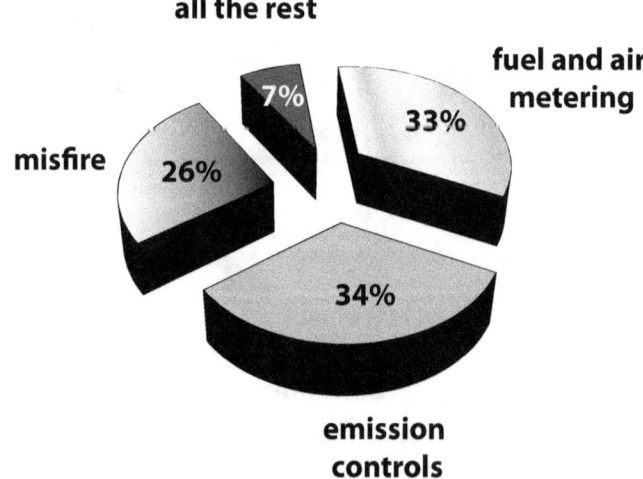

all the rest 7%
fuel and air metering 33%
misfire 26%
emission controls 34%

# Passing the Emissions Test

## Pass the Test the First Time

Passing a scan tool emissions test the first time through is a big plus. In some states, if your vehicle fails, **the catalyst monitor will have to be one of the completed monitors** when the vehicle is retested.

Here's why this is important.

After DTCs are erased, some vehicles will take their sweet time running the catalyst monitor. My own car, which has been used for many experiments, still passes the emissions test, every time. But if I erase its DTCs, it may take **two weeks** to complete the cat monitor—even if I use the OEM drive cycle and follow it to the letter.

If your license plates expire tomorrow, this is a big deal.

In our home state, a vehicle with this readiness list will not be turned away at the test center on a first test. Assuming it has no other problems, it has only one incomplete monitor, and will pass the scan tool emissions test.

However, this vehicle would not pass a **retest**. It may have only one incomplete monitor, but since it is the **catalyst monitor**, it will have to be driven until the catalyst monitor runs to completion. Play it safe; make sure your car passes the first time!

## Passing the Emissions Test Review

- Erasing DTCs is not always a good idea. After the repair is made, it may be preferable to drive the vehicle and let the PCM turn off the MIL. This does not reset monitors.

- A generic drive cycle commonly runs most monitors.

- Some monitors will not run during the generic drive cycle. Look up the drive cycle description in a manual and follow it to the letter, in these instances.

- Monitors run only when specified operating conditions are present. These conditions are called *enabling criteria*.

- The vast majority of vehicle failures resulting in DTCs are related to misfire, improper air/fuel ratio, and failed emission controls. A tiny percentage are attributable to transmissions and vehicle computers.

- Many vehicle failures are still corrected by routine maintenance.

# SECTION TWO
# ADVANCED
# TROUBLESHOOTING

# What's In Section Two?

Welcome to **Section Two** of the book,  **Advanced Troubleshooting.**

In this section, we look at three topics that require our special attention:

**Mode 6** - While it is one of the original OBD II test modes, Mode 6 has been alternately praised as a magic bullet cure-all for all OBD II failures, or blasted as an arcane waste of time. We believe the truth lies somewhere in between these two extremes.

When used regularly, Mode 6 can provide us with useful information about the test status of various monitors, and also offers additional information in some makes about engine misfire and oxygen sensor tests.

We will show you how to avoid common pitfalls, and encourage you to use Mode 6 often enough to make it a valued addition to your diagnostic tool box.

---

**NOx** - There has already been ample mention of NOx in Section One. But NOx poses a special set of problems that require closer scrutiny.

The problem with NOx is that it is a common side effect of HC and CO repairs. It is also associated with overdue vehicle maintenance that results in elevated combustion temperatures.

The NOx section addresses these issues in greater detail, and offers common cures.

---

**Diesel Emissions** - This is a relatively new subset of the emission control mandate. Our overview introduces substantive changes in diesel fuel composition, and a whole new hardware suite designed to control the old emission gases, plus a new emissions danger: tiny soot particles that pose a particularly insidious threat to our health.

In the space we have, we can only provide  a broad summation, but we hope it gets you started in a more comprehensive study of diesel emission strategies and repairs.

# MODE 6

# 12

# Mode 6

## What Is Mode 6?

Mode 6 is one of the ten OBD II modes. It is an advanced diagnostic tool, a part of the generic test interface. (See page 287.)

- Mode 6 lets us review test results of certain monitors, most often noncontinuous monitors, although Mode 6 may also display selective continuous monitor and oxygen sensor test data.

- Mode 6 is a generic tool, but the test results it displays are all manufacturer-specific.

- Mode 6 data vary by monitor, make, model and even model-year.

- Raw OBD data gathered with Mode 6 must be "translated" before we can use it. Some scan tools do this for us; others make us look up the specifics in a repair manual.

### Mode 6 Translation

Think of it this way: Mode 6 is a transmitter, like a transoceanic radio. It broadcasts data from a place inside the PCM that speaks "computer." Like broadcasts from foreign countries, we must translate the content to understand it.

To view Mode 6 data, we need:

- A scan tool that displays Mode 6 (not all do).
- A vehicle that supports Mode 6 (not all do).
- A translator that converts raw Mode 6 data into useful information. We can view data in an understandable format from a scan tool with a built in "translator," or we can translate the data manually, using OEM repair information.

The second option is tedious, time consuming, and difficult. In fact, we'll go so far as to say that it is impractical for most of us, and will not be covered here.

# Mode 6

## What Does Mode 6 Look Like?

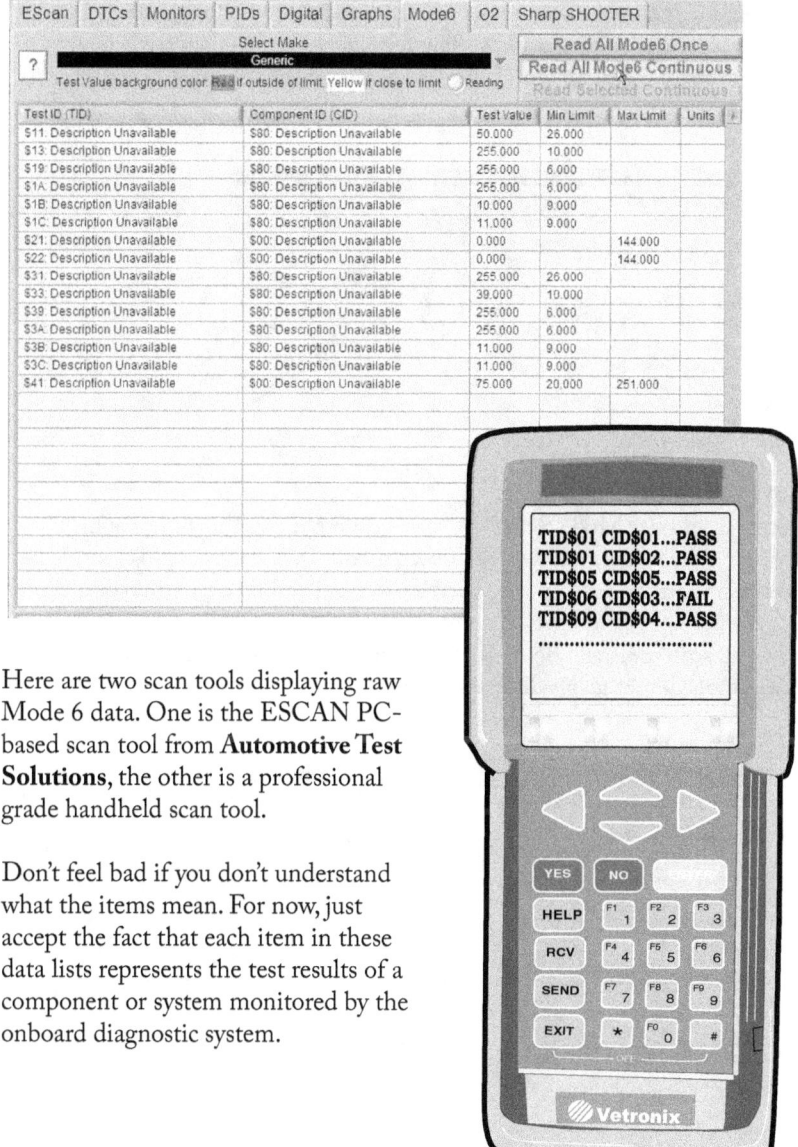

| Test ID (TID) | Component ID (CID) | Test Value | Min Limit | Max Limit | Units |
|---|---|---|---|---|---|
| $11: Description Unavailable | $80: Description Unavailable | 50.000 | 26.000 | | |
| $13: Description Unavailable | $80: Description Unavailable | 255.000 | 10.000 | | |
| $19: Description Unavailable | $80: Description Unavailable | 255.000 | 6.000 | | |
| $1A: Description Unavailable | $80: Description Unavailable | 255.000 | 6.000 | | |
| $1B: Description Unavailable | $80: Description Unavailable | 10.000 | 9.000 | | |
| $1C: Description Unavailable | $80: Description Unavailable | 11.000 | 9.000 | | |
| $21: Description Unavailable | $00: Description Unavailable | 0.000 | | 144.000 | |
| $22: Description Unavailable | $00: Description Unavailable | 0.000 | | 144.000 | |
| $31: Description Unavailable | $80: Description Unavailable | 255.000 | 26.000 | | |
| $33: Description Unavailable | $80: Description Unavailable | 39.000 | 10.000 | | |
| $39: Description Unavailable | $80: Description Unavailable | 255.000 | 6.000 | | |
| $3A: Description Unavailable | $80: Description Unavailable | 255.000 | 6.000 | | |
| $3B: Description Unavailable | $80: Description Unavailable | 11.000 | 9.000 | | |
| $3C: Description Unavailable | $80: Description Unavailable | 11.000 | 9.000 | | |
| $41: Description Unavailable | $00: Description Unavailable | 75.000 | 20.000 | 251.000 | |

Here are two scan tools displaying raw Mode 6 data. One is the ESCAN PC-based scan tool from **Automotive Test Solutions,** the other is a professional grade handheld scan tool.

Don't feel bad if you don't understand what the items mean. For now, just accept the fact that each item in these data lists represents the test results of a component or system monitored by the onboard diagnostic system.

# Mode 6

## Uninterpreted Data

This screen shows raw Mode 6 data. The arrow points to the word "generic." In generic mode, the scan tool does not translate the raw data. Individual tests are labeled with numbers. Raw test values are not converted to common units of measurement. If we had access to the vehicle maker's secret decoder sheet, we could sit down and make the conversions, manually.

| Test ID (TID) | Component | Test Value | Min Limit | Max Limit | U |
|---|---|---|---|---|---|
| S05: Description Unavailable | S01: Descri | 9599.000 | | 65535.000 | |
| S05: Description Unavailable | S02: Descri | 11200.000 | | 65535.000 | |
| S05: Description Unavailable | S03: Descripti | .000 | | 65535.000 | |
| S05: Description Unavailable | S04: Description | 00 | | 65535.000 | |
| S05: Description Unavailable | S05: Description Unavailable | 00 | | 6720.000 | |
| S05: Description Unavailable | S06: Description Unavailable | .000 | | 6080.000 | |
| S05: Description Unavailable | S87: Description Unavailable | 5.000 | 65.000 | | |
| S05: Description Unavailable | S88: Description Unavailable | 212.000 | 50.000 | | |
| S06: Description Unavailable | S41: Description Unavailable | 74.000 | | 390.000 | |

## Interpreted Data

Here is another Mode 6 display from a different car. This time, we have selected the vehicle from a drop down menu. This scan tool looks inside its resident library, looks up the Chrysler section, and then interprets the raw data into words and units of common measurement that we can understand. Not all scan tools will do this. Some display only raw Mode 6 data. Some scan tools do not display Mode 6 at all.

| Test ID (TID) | Component ID (CID) | | Min Limit | Max Limit | Units |
|---|---|---|---|---|---|
| S11: O2 Sensor Monitor | S80: B1S1 Half-Cycle Counter | | 000 | | count |
| S13: O2 Sensor Monitor | S80: B1S1 Big Slope Counter | | | | count |
| S19: O2 Heater Monitor | S80: B1S1 Hot Trend Counter | | | | count |
| S1A: O2 Heater Monitor | S80: B1S2 Hot Trend Counter | | | | count |
| S1B: O2 Heater Monitor | S80: B1S1 Heater Voltage | 200. | | | mv |
| S1C: O2 Heater Monitor | S80: B1S2 Heater Voltage | 200.000 | | | mv |
| S21: Catalyst System Monitor | S00: Bank 1 Switch Ratio Frequency | 0.000 | | 6.160 | % |
| S22: Catalyst System Monitor | S00: Bank 2 Switch Ratio Frequency | 2.340 | | 56.160 | % |
| S31: O2 Sensor Monitor | S80: B2S1 Half-Cycle Counter | 255.000 | 26.000 | | count |
| S33: O2 Sensor Monitor | S80: B2/S1 Big Slope Counter | 86.000 | 10.000 | | count |
| S39: O2 Heater Monitor | S80: B2S1 Hot Trend Counter | 255.000 | 6.000 | | count |
| S3A: O2 Heater Monitor | S80: B2S2 Hot Trend Counter | 255.000 | 6.000 | | count |
| S3B: O2 Heater Monitor | S80: B2S1 Heater Voltage | 220.000 | 180.000 | | mv |
| S3C: O2 Heater Monitor | S80: B2S2 Heater Voltage | 220.000 | 180.000 | | mv |

The Auto Emissions Bible

# Mode 6

## What's Real, What Isn't

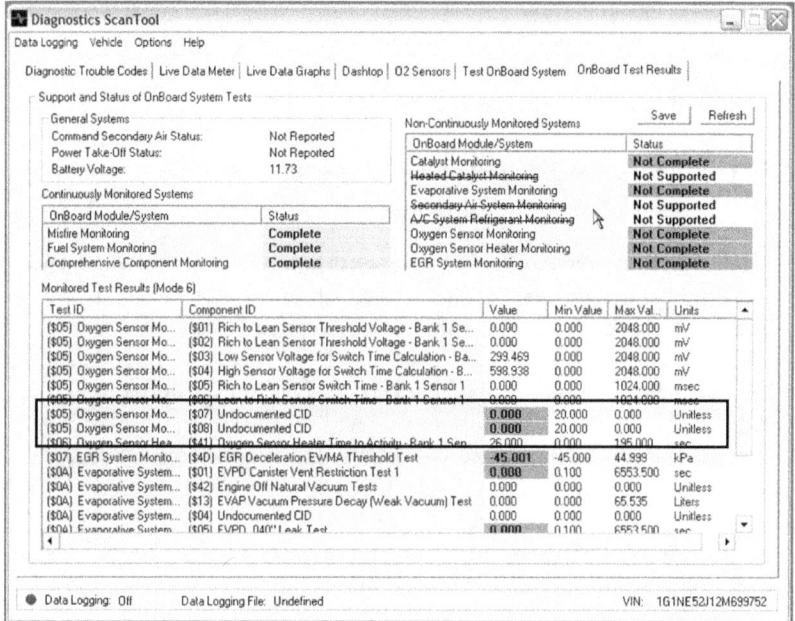

Here is the **Auto Enginuity** PC-based scan tool displaying Mode 6 data from a Chevy Malibu. Most of the parameters and values are identified in words and measurement scales we can identify.

But two data parameters are listed as "undocumented." Huh?

Sometimes an OEM uses a "master list" for all vehicles in a given model year. The master list contains all possible Mode 6 parameters available, whether or not they are used in a given vehicle. If Mode 6 items are available but not used, they remain in the list as placeholders. That's what we are seeing here: empty boxes marked "undocumented."

We have just begun, and already we have variations on a theme. We are about to see other inconsistencies in Mode 6 real world application, so stay tuned.

# Mode 6

## Mode 6 Operational Overview

OBD II is an ongoing series of test procedures that check the vehicle for failures that MIGHT result in increased vehicle emissions. Each test has a set of PASS/FAIL standards. Each monitor may have several tests.

Here is how Mode 6 is supposed to work:

- **To view test results, the scan tool sends a request to the PCM for Mode 6 data.**

- **Mode 6 lists the systems and components being tested.** It also displays actual test values and the test limits to which they are compared. Raw data lists won't include labels or common measurement units.

- **Vehicle manufacturers assign Test IDs (TIDs) and Component IDs (CIDs) to systems and components.** Test limits for many of these components and systems are found in Mode 6. This makes it a great source for test limits that may not be available in poorly written DTC descriptions.

- **The most recent Mode 6 test values *should* be retained in the PCM, even over multiple key cycles.** Mode 6 data *should* be replaced *only* by more recent test data. (In real application, this is not always the case.)

- **Mode 6 data are manufacturer-specific.** This includes system and component labeling, and test standards.

- **Pass/Fail standards are referred to as test limits.** To pass a test, a component test value must be:

- below a maximum.
- above a minimum.
- between a minimum and a maximum.

## Mode 6 Translated

Let's get this TID and CID thing clarified.

| Test ID (TID) | Component ID (CID) | Test Value | Min Limit | Max Limit | Units |
|---|---|---|---|---|---|
| $05: O2 Sensor Monitors and Constants | $01: B1S1 Rich to Lean Threshold | 0.000 | | 2047.969 | mV |
| $05: O2 Sensor Monitors and Constants | $02: B1S1 Lean to Rich Threshold | 0.000 | | 2047.969 | mV |
| $05: O2 Sensor Monitors and Constants | $03: B1S1 Low Switch Time Calculation | 299.469 | | 2047.969 | mV |
| $05: O2 Sensor Monitors and Constants | $04: B1S1 High Switch Time Calculation | 598.937 | | 2047.969 | mV |
| $05: O2 Sensor Monitors and Constants | $05: B1S1 Rich to Lean Switch Time | 25.992 | | 104.966 | msec |
| $05: O2 Sensor Monitors and Constants | $06: B1S1 Lean to Rich Switch Time | 21.993 | | 94.970 | msec |
| $05: O2 Sensor Monitors and Constants | $87: B1S1 Rich to Lean Switches | 238.000 | 65.000 | | sw |
| $05: O2 Sensor Monitors and Constants | $88: B1S1 Lean to Rich Switches | 242.000 | 50.000 | | sw |
| $06: O2 Sensor Heater Monitor | $41: B1S2 Heater Time to Activity | 67.000 | | 355.000 | sec |
| $06: O2 Sensor Heater Monitor | $35: B1S1 Heater Time to Activity | 26.000 | | 87.000 | sec |
| $07: Exhaust Gas Recirc Sys Monitor | $4D: EGR decel test | -5.067 | | 1.699 | kPa |
| $0A: EVAP Monitor #2 (.020 Leak) | $01: EVPD canister vent restriction test 1 | 0.000 | | 15.000 | sec |
| $0A: EVAP Monitor #2 (.020 Leak) | $42: EVPD canister vent restriction test 2 | 0.600 | | 10.000 | inH2O |
| $0A: EVAP Monitor #2 (.020 Leak) | $03: EVPD weak vacuum test | 0.000 | | 5.000 | sec |
| $0A: EVAP Monitor #2 (.020 Leak) | $84: EVPD weak vacuum followup test | 0.000 | 0.000 | | sec |
| $0A: EVAP Monitor #2 (.020 Leak) | $05: EVPD .040" leak test | 0.042 | | 0.035 | inches |
| $0A: EVAP Monitor #2 (.020 Leak) | $87: EVPD purge pass test | 50.000 | 50.000 | | sec |
| $0A: EVAP Monitor #2 (.020 Leak) | $48: EVPD purge vacuum fail test | 0.700 | | 5.000 | inH2O |
| $0C: Catalyst Efficiency Monitor | $60: Bank 1 Catalyst Test OSC | -9.488 | | 0.281 | sec |

This screen capture of Mode 6 data shows systems (listed under TIDs), tests within each system (listed under CIDs), test limits, test results, and units of measurement.

- Note the two test and component columns: TIDs (for Test IDs)and CIDs (for Component IDs).

- In this GM vehicle, TIDs indicate a **group** of sensors or a system; things like oxygen sensors, EGR valves, and the evaporative emission system.

- CIDs represent a specific **component test** within each TID group. For example, there are 10 different oxygen sensor tests listed. There is one EGR test, followed by 7 EVAP tests, with one lonely catalyst test rounding out the list.

- In the columns to the right, we see actual test results and the test standard limits, either minimum or maximum.

- The scan tool has gone to the trouble of highlighting a failed test with a dark highlight to make it stand out as a failure. Clearly, a scan tool with support features like a translator and error highlighting make life much easier for the diagnostician.

# Mode 6

## TIDs and CIDs

Here's a graphic layout showing TIDs and CIDs at work in a sample vehicle. This OEM has 10 separate tests for the oxygen sensors; 7 for the EVAP system, and one each for the catalyst and EGR.

For this vehicle, TIDs are systems. CIDs are components.

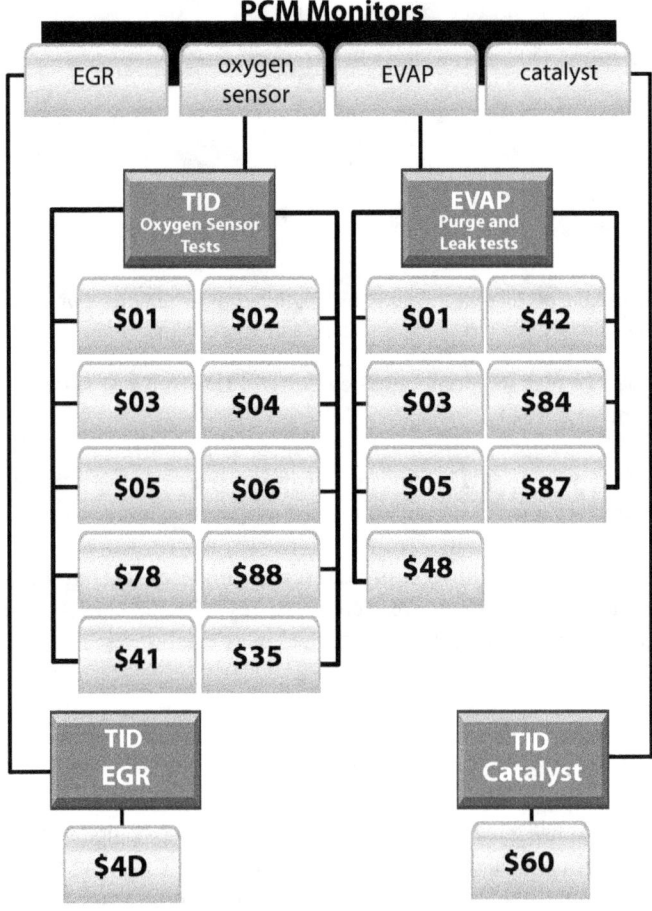

Now the bad news: TIDs and CIDs mean different things to different vehicle makers. On the next page, we'll show you an example of a vehicle where the TID indicates the type of test, not the system being tested. Engineers....

# Mode 6

## Interpreting a Mode 6 Minimum Test

The screen on our scan tool displays TID 16 on a 2003 Chrysler. But TIDs and CIDs mean different things on this vehicle that they did in the previous example.

- **TID 16 is the O₂ sensor 1/2 Rich (HIGH) voltage test.** This test determines if the bank one downstream oxygen sensor voltage crosses a voltage threshold.

- **The CID (Component ID) is 80.** (Compare that to the CID on the opposite page.) In this Chrysler computer, the CID does not indicate a component! Here, it indicates whether the test type is a minimum or a maximum. The number 80 indicates a test with a **minimum** value. When a minimum test limit is given, the test value must **equal or be *more* than** the limit to pass.

- **Factory reference material for this vehicle tells us that the raw data multiplied by 20 is equal to millivoltage.** To pass this test, the oxygen sensor voltage must reach or exceed 760 millivolts (38 x 20=760mV). As soon as the test limit is reached, the test is a pass.

The actual voltage reached 760mV. The test is a pass.

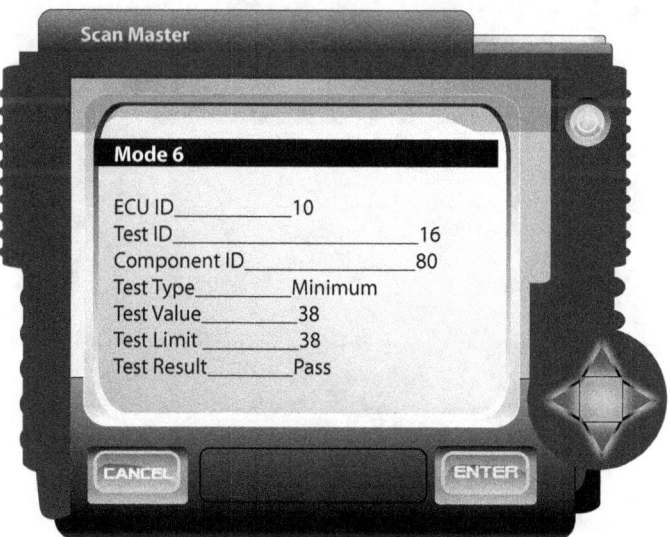

For a **MIN** test to pass, the measured value must be **greater than or equal to the limit**.

# Mode 6

## Interpreting a Mode 6 Maximum Test

The screen on our scan tool displays TID 17 on a 2003 Chrysler.

- **Chrysler factory reference material tells us that TID 17 is the O$_2$ sensor 1/2 Lean (LOW) voltage test.** This test determines if the downstream oxygen sensor voltage can reach a low voltage threshold.

- **The CID (Component ID) is 00.** In this Chrysler the CID does not indicate a component. Instead, it indicates whether the test type is a minimum or a maximum. Here, a CID of 00 indicates a **maximum** test. On our Chrysler, when a maximum test limit is given, the test value must be **equal to or less** than the limit to pass.

- **Factory reference material for this vehicle tells us that the raw data multiplied by 20 equals millivoltage.** To pass this test, the oxygen sensor voltage must be equal to or less than 580 millivolts (29 x 20=580mV). As soon as the test limit is reached, the test is a pass. (Converting to millivolts is useful when we want to double check a circuit with a meter.)

- The actual voltage reached was 580mV. The test is a pass.

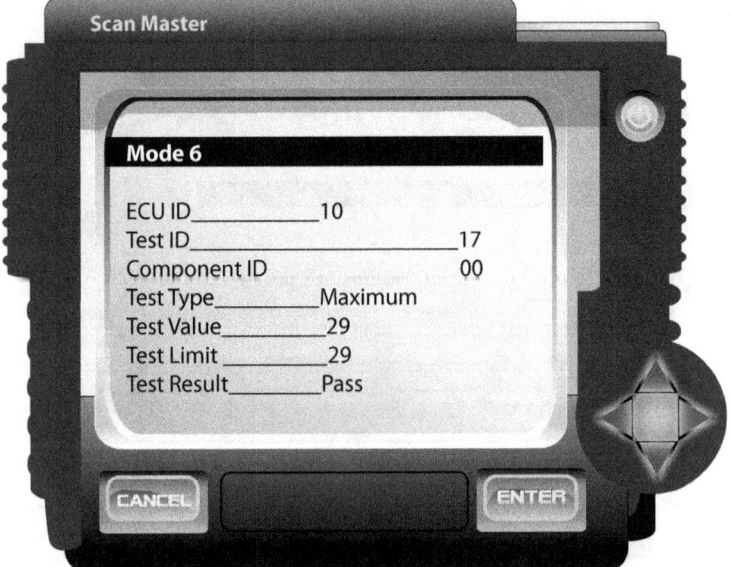

For a **MAX** test to pass, the measured value must be **less than or equal to the limit**.

The Auto Emissions Bible

## Not Too High, Not Too Low: Just Right

Some Mode 6 test limits combine a high and low test value. To pass, the test value must be **less than the maximum** AND **greater than the minimum**.

Here is an example of an oxygen sensor heater test.

Note that this test has a both a minimum and a maximum test limit; for the monitor to record the test as a pass, the test value must fall *between* them. If the test value is outside this range, the monitor records a "fail."

| TID | CID | Test Value | MIN | MAX | Unit | Pass/Fail |
|---|---|---|---|---|---|---|
| Oxygen Sensor Monitoring | MIN/MAX O2 Heater Current | 1.164 | 0.465 | 3.0 | A | ? |

The image below helps illustrate the concept.

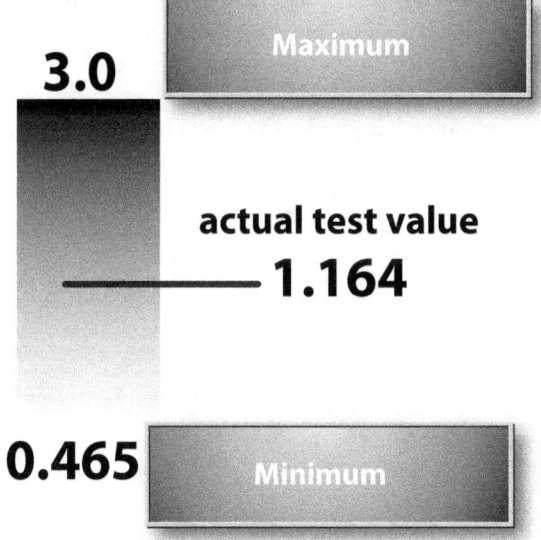

### For this test:

- Test results **above** 3.0 are recorded as a **fail** by the monitor.
- Test results **below** 0.465 are recorded as a **fail** by the monitor.
- Our test value falls safely between the high and low limits; is a **pass**. Be suspicious of test values that are very close to a limit, a common warning of an impending problem.

**Note:** Some tests end as soon as they pass. When this happens, you will not know how much "better" the component or system would have scored had the test continued.

# Mode 6

## Mode 6 Limitations

Mode 6 is not perfect.

- **Many scan tool interfaces don't support Mode 6.** Don't expect to see Mode 6 in many older scan tools or in inexpensive scan tools with a limited feature set.

- **Ideally, the PCM should store the most current test results in memory until they are overwritten by more current test results.** Note that we said, "ideally." Sorry, but some vehicles will reset Mode 6 data each time the ignition is switched off. That means you'll need to access data from the current driving cycle *before* turning off the key. Some vehicles will not update test results until they are shut off and restarted.

- **Unless your scan tool does it for you, you will need manufacturer-specific Mode 6 conversion data to interpret raw Mode 6 data.** One of the most frequently asked questions is, "Isn't there a master list we can refer to?" Not yet, if ever. Prepare yourself to do some research if your scan tool doesn't act as an interpreter, converting raw data into useful information.

| | | | | | |
|---|---|---|---|---|---|
| $01: A/F Sensor Monitor (Bank 1) | $89: Description Unavailable | 875.000 | 130.000 | 65535.000 | |
| $02: Oxygen Sensor Monitor (Bank 1) | $9D: Description Unavailable | 0.048 | 0.000 | 0.265 | V |
| $02: Oxygen Sensor Monitor (Bank 1) | $9E: Description Unavailable | 0.157 | 0.050 | 65.535 | V |
| $21: Catalyst System Monitor (Bank 1) | $A1: Catalyst Test | 0.000 | 0.000 | 3.000 | V |
| $31: EGR System Monitor | $D0: EGR Valve Target/Lift Difference | 0.006 | 0.000 | 0.040 | inch |
| $31: EGR System Monitor | $D1: EGR Valve Lift Quantity | 0.083 | 0.006 | 1.999 | inch |
| $31: EGR System Monitor | $D2: EGR Valve Lift Test | 0.000 | 0.000 | 0.000 | |
| $31: EGR System Monitor | $D3: EGR Monitor Using AF Sensor | 15646.000 | 0.000 | 36045.000 | |
| $3A: EVAP System Monitor | $BA: Fuel Tank Atmospheric Pressure | 41.000 | 0.000 | 1200.000 | sec |
| $3C: EVAP System Monitor | $B4: Fuel Tank Pressure Increase | 0.605 | 0.000 | 1.999 | |
| $3C: EVAP System Monitor | $B5: Fuel Tank Pressure Increase | 0.000 | 0.000 | 0.000 | |
| $3C: EVAP System Monitor | $B6: Fuel Tank Atmospheric Pressure | 0.000 | 0.000 | 0.000 | sec |
| $3D: EVAP System Monitor | $B9: EVAP Canister Purge Test Ratio | 100.006 | 30.001 | 100.006 | % |
| $A2: Misfire Monitor Cylinder 1 | $0B: Cylinder #1 Misfire Rate | 0.000 | 0.000 | 65535.000 | Counts |
| $A2: Misfire Monitor Cylinder 1 | $0C: Cylinder #1 Misfire Rate | 0.000 | 0.000 | 65535.000 | Counts |
| $A3: Misfire Monitor Cylinder 2 | $0B: Cylinder #2 Misfire Rate | 0.000 | 0.000 | 65535.000 | Counts |
| $A3: Misfire Monitor Cylinder 2 | $0C: Cylinder #2 Misfire Rate | 4.000 | 0.000 | 65535.000 | Counts |
| $A4: Misfire Monitor Cylinder 3 | $0B: Cylinder #3 Misfire Rate | 0.000 | 0.000 | 65535.000 | Counts |
| $A4: Misfire Monitor Cylinder 3 | $0C: Cylinder #3 Misfire Rate | 0.000 | 0.000 | 65535.000 | Counts |
| $A5: Misfire Monitor Cylinder 4 | $0B: Cylinder #4 Misfire Rate | 0.000 | 0.000 | 65535.000 | Counts |
| $A5: Misfire Monitor Cylinder 4 | $0C: Cylinder #4 Misfire Rate | 0.000 | 0.000 | 65535.000 | Counts |

Mode 6 data from a 2011 Honda Fit. Note the presence of misfire data.

It is NOT always clear what types of information you'll find when you ask for Mode 6 data. In addition to regular Mode 6 parameters, a vehicle maker may decide to display heated oxygen sensor test data in Mode 6, instead of using Mode 5 (the OBD II designated mode for oxygen sensor test data). Another exception is a decision to display misfire data in Mode 6, even though the Misfire Monitor is a continuous, not a non-continuous monitor. (Actually, this is a nice feature if you know it's there.)

The Auto Emissions Bible

# Mode 6

## Mode 6 - Use With Caution

Here's how Mode 6 can drive you crazy!

- **Somebody just erased DTCs with a scan tool or replaced the car battery.** These actions also erased all Mode 6 data. You hook up your scan tool and the Mode 6 data screen you see has a lot of missing and illogical data. These data will be partially or totally useless. Not much help here.

- **After a PCM reset, some vehicles will not report Mode 6 data until the monitors have run to completion.** Others will display Mode 6 data screens filled with meaningless placeholder numbers until the tests run and the real test results become available. Be careful.

- **ALWAYS look at several data screens and compare them.** Take before and after shots. Assuming DTCs were not erased, take a pre-repair screen capture before you fix or change anything. If the data are useful, use them to help you repair the vehicle. After the repairs are completed, drive the vehicle (without erasing DTCs) and grab fresh data for comparison.

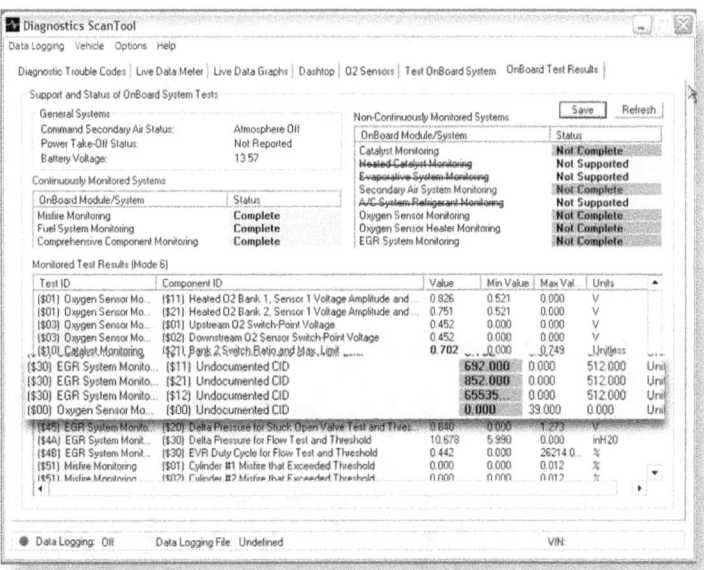

This Mode 6 display on an **Auto Enginuity** PC-based scan tool is displaying some weird, random numbers gathered right after DTCs are erased. See the items listed as "undocumented"? The scan tool correctly highlights these items as questionable.

# Mode 6

## Crazy Data

Lets look at sample Mode 6 data from a 1997 Lincoln Continental. This vehicle has no DTCs, the MIL is commanded OFF, and the screen capture shown below tells us that all monitors are complete. This vehicle would pass an emissions test.

Let's erase DTCs and see what happens to Mode 6 data.

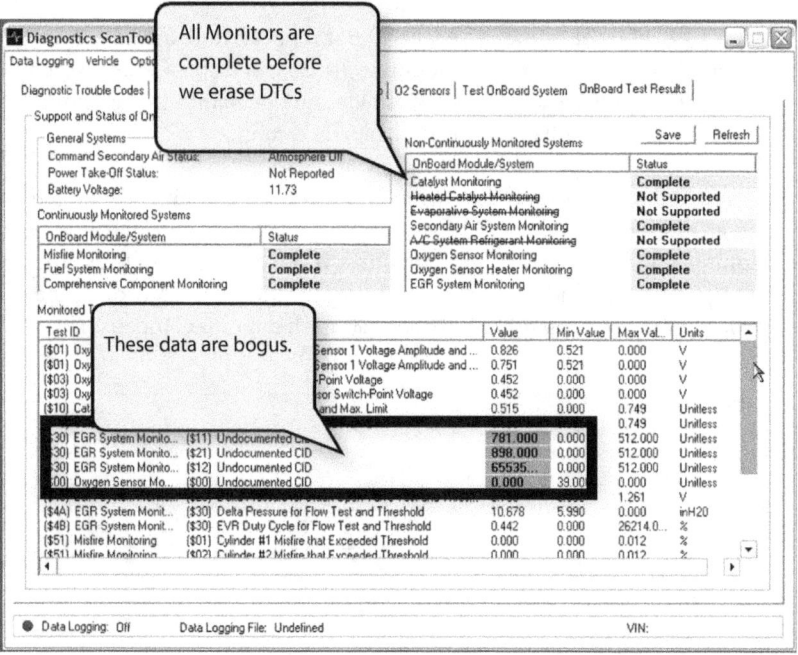

This screen is a real example of why it is important to use Mode 6 frequently, if you plan to use it at all. Experience is the best teacher.

Even though this vehicle would pass a scan tool emissions test, several Mode 6 data lines have been highlighted by the scan tool and labeled as *undocumented*. But there is nothing to fix. These parameters are not used in this vehicle.

The correct response is to ignore the phantom parameters.

# Mode 6

## Crazier Data

Let's experiment. Here's Mode 6 data from the same 1997 Lincoln **after we erase DTCs.**

The engine has been allowed to idle for several minutes.

- Erasing DTCs has erased all Mode 6 data.
- Non-continuous monitors have been reset to "not complete."
- The same "bogus" parameters highlighted in the previous screen are still marked as suspicious.
- The test values for the catalyst switching frequency ratio are *identical* for both banks, which is illogical. Our Ford PCM has plugged in "placeholder" data for some parameters until real data are collected. We need to drive this vehicle until the tests run again, and view a fresh Mode 6 screen after it is updated with real test results.

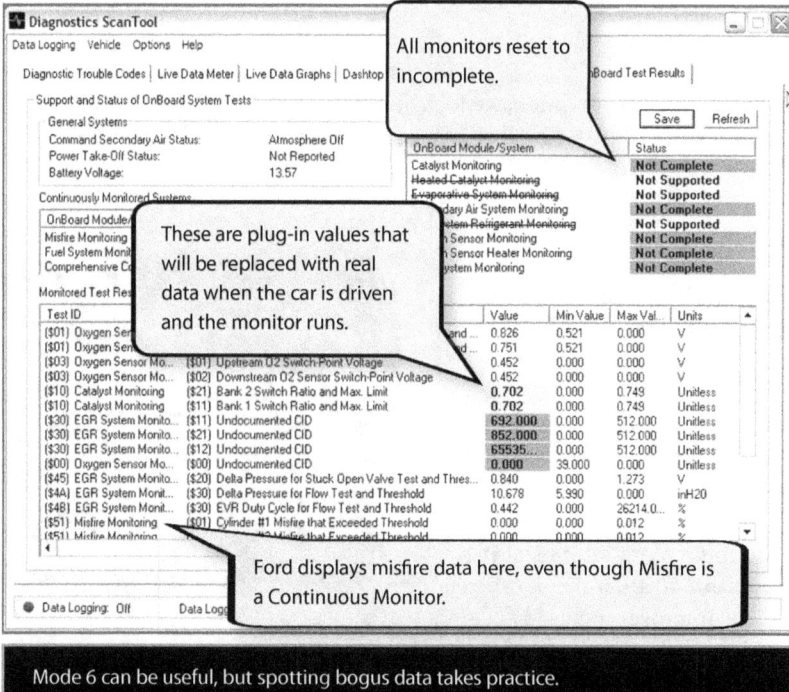

All monitors reset to incomplete.

These are plug-in values that will be replaced with real data when the car is driven and the monitor runs.

Ford displays misfire data here, even though Misfire is a Continuous Monitor.

Mode 6 can be useful, but spotting bogus data takes practice.

Play it safe: view multiple Mode 6 snapshots and compare them. You'll never know a "bad" car unless you get into the habit of looking at GOOD cars for reference.

# Mode 6

## Mode 6 Case Study

**Vehicle:** 2002 Mass Market Sedan
**Engine:** 2.0L
**Odometer:** 97,000 miles
**DTC(s):** P0420 (Low Catalyst Efficiency), P0301, P0305 (Misfires Cylinders 1 and 5)
**MIL:** ON
**Test Limit:** Mode 6 data indicate that the test limit for the catalyst efficiency test is 0.810.

The MIL is on and a DTC P0420 has been stored indicating low catalyst efficiency. The vehicle has been severely neglected and misfires at all engine speeds. Leaking secondary insulation and worn spark plugs are the causes for the multiple misfires. We correct the misfire condition(s) first, since misfire damages catalysts.

To verify catalyst condition, we look at Mode 6 data. The most recently reported test results for catalyst efficiency indicate a test value of 0.790. This is a pass (less than 0.810).

Should we replace the catalyst?

Probably. Even though the catalyst has passed its most recent test, it has done so by a razor thin margin. It is obviously damaged and is failing its efficiency tests often enough to store a DTC and turn on the MIL.

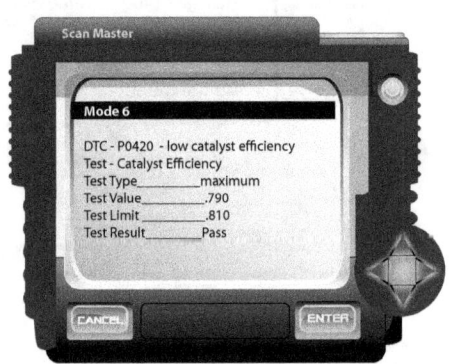

(You could retest after repairing the misfire to see if the next batch of tests improve, or erase DTCs and let the monitor run again, checking the Mode 6 updates.)

Compare this scan tool display to the graphed data on pages 182-183. We can view test data in a number of ways, and cross reference results to improve our diagnoses.

# Mode 6

A new OEM-equivalent catalyst is installed. Then we drive the vehicle to get the catalyst hot. When the catalyst monitor runs to completion, Mode 6 catalyst efficiency data change dramatically. The switching frequency drops to 0.070. Quite an improvement!

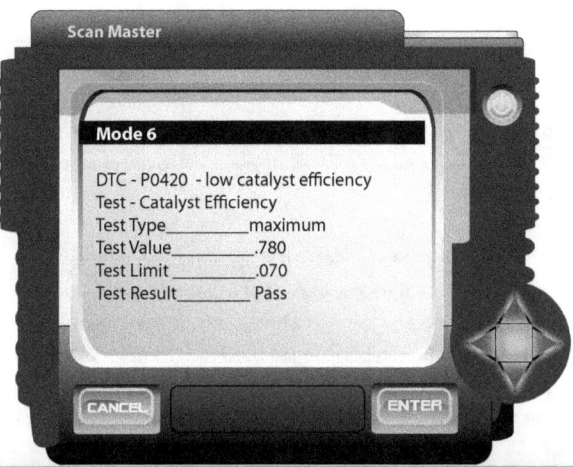

Scan Master

**Mode 6**

DTC - P0420 - low catalyst efficiency
Test - Catalyst Efficiency
Test Type_____maximum
Test Value_____.780
Test Limit _____.070
Test Result_____ Pass

CANCEL          ENTER

Points to consider in this exercise:

• The lower the catalyst switching ratio, the better. As the catalyst degrades, the switching rate of the downstream oxygen sensor increases.

• This test limit is a maximum.

• Using an aftermarket replacement that does not meet OEM standards is a gamble. Catalysts with smaller capacity (smaller catalyst bed/less oxygen storage) than the original may test poorly compared to an OEM-equivalent replacement. Even if they satisfy the catalyst monitor originally, they will die sooner than a replacement with OEM capacity.

Misfires kill catalysts. Any time you see a low catalyst efficiency DTC coupled with misfire codes, correct the misfire problem(s) before replacing the old catalyst, or you'll kill the new one too!

## Mode 6 - Review

- Mode 6 can help us verify our repairs faster. After we complete a repair, we really want to know if the MIL will stay off. That IS the point of this exercise, right?

- Mode 6 can provide repair verification in less time than it takes to complete all monitors. (Imagine looking at a teacher's grade book for individual quizzes, before he issues a final grade.) We get feedback about the most recent status of test strategies used with complex EVAP and Catalyst monitors. Test results give an early indication of repair success. This saves time and reduces the likelihood of a MIL-related comeback.

- Sometimes a vehicle experiences a driveability symptom that does not store a DTC. The condition may worsen until a DTC is finally stored. Do we need to wait until there is a DTC stored to guide us?

- Some vehicles may misfire with no MIL illumination or DTC. Mode 6 can identify misfire in individual cylinders in some vehicles.

- If you're planning on using Mode 6 only on problem cars, you'll probably be disappointed with the results. Inconsistencies in Mode 6 application can cause a lot of wasted effort for the occasional Mode 6 user. Successful Mode 6 practitioners combine data with regular practice to get the most from this unique diagnostic tool.

# NOx 13

# NOx

## NOx as Nemesis

This chapter is dedicated to solving NOx problems. NOx can lie dormant in rich running engines but rear its ugly head as soon as you conquer HC and CO. Repairs of rich running engines often result in leaner, hotter mixtures that contribute to NOx formation starting at 2500°F. (See page 25.)

With temperature so critical to NOx formation, the cooling system must be treated as an emissions control device. Failure to remove heat from the cylinders creates hot spots that breed NOx.

Check engine cooling system condition and operation whenever you encounter a NOx failure. Just a few degrees of extra coolant temp can make a big difference.

At a minimum, perform a thorough visual inspection of all belts, hoses, and the water pump/drive belt. Monitor the engine coolant temperature PID on your scan tool while driving at moderate engine speeds with the heater fully on HOT and heater fan on HIGH. Look for high temperatures or big temperature swings that may indicate a faulty thermostat.

Then turn the heater off. A spike in the ECT PID indicates restricted coolant passages or reduced air flow.

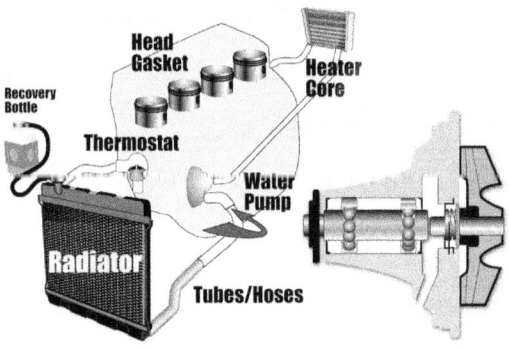

Neglected cooling systems, sticking thermostats, and weak water pumps are prime suspects when $NO_x$ failures occur.

## Preventing NOx

- **Remove accumulations of debris from the radiator and AC condenser that restrict air flow.** Be careful not to bend or damage tubing fins as you blow away debris. Bent or missing fins restrict air flow and reduce heat transfer.

- **Pay particular attention to the area between the condenser and radiator.** It can turn into a rat's nest of paper, leaves, and debris. Be sure to replace all torn or missing seals around and between the radiator-condenser assembly. These seals keep debris out and push more air through the condenser/radiator.

- **Replace/repair any body air dams that direct air flow.** Closed-front aerodynamic vehicle designs require these dams to direct air through the radiator for maximum cooling.

- **Flow test the radiator if the engine overheats under load or when the AC is on.** Use an infrared temperature gun to check for localized hot spots that indicate reduced coolant flow.

- **Check the accessory belt condition, tension, and alignment.** In vehicles with serpentine belts and automatic belt tensioners, make sure the belt is the correct length for the application. Changes in overall belt length of less than one inch can result in belt slippage.

- **Beware of underdrive pulleys** that slow belt driven accessories to add horsepower. These may turn the water pump too slowly to properly circulate coolant.

> Point and shoot temperature probes take the guesswork out of cooling system problems that contribute to $NO_x$ formation.

# NOx

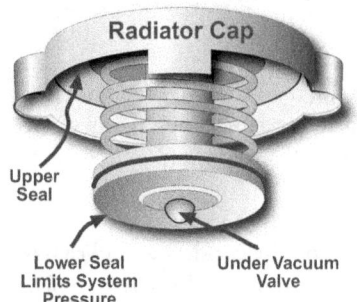

Radiator Cap

Upper Seal

Lower Seal
Limits System
Pressure

Under Vacuum
Valve

## Radiator Caps

What does a humble radiator cap have to do with NOx? Plenty.

A faulty radiator cap can leave the radiator low on coolant, leading to overheating and an increase in NOx emissions.

Engine coolant expands when hot. The extra pressure this creates opens the radiator cap seal, venting coolant to a hose attached to an overflow bottle. The coolant level in the overflow rises when the engine and its coolant are hot, as a result.

Later, when the car is parked, engine off, coolant inside the radiator cools and contracts. This creates a vacuum that pulls coolant back from the overflow to the radiator.

Coolant transfer, both hot and cold, keeps the cooling system filled with liquid. If the radiator cap fails, the coolant bottle will overfill, creating an air pocket in the radiator that leads to overheating.

### Hot- Pressurized System

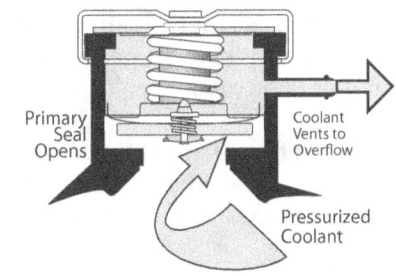

Primary
Seal
Opens

Coolant
Vents to
Overflow

Pressurized
Coolant

### Cold System - Low Pressure

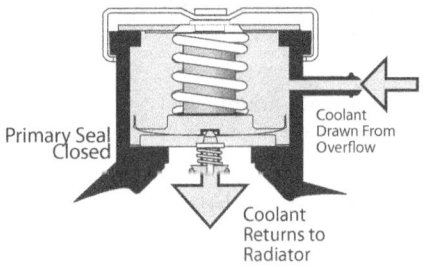

Primary Seal
Closed

Coolant
Drawn From
Overflow

Coolant
Returns to
Radiator

Test and, if necessary, replace a defective radiator cap. Do not assume that the radiator is full just because the overflow bottle has coolant inside.

## Cooling Fan Codes

Cooling is so important to NOx control, that a DTC may be stored if a fault is detected in a cooling fan control circuit. We recently fixed a wiring fault in a car that had stored a DTC: **P0481 - Cooling Fan 2 Control Circuit Malfunction**.

The problem was traced to a control wire from the PCM to the number 2 cooling fan relay. Battery vapors had eaten through the relay ground wire. Fan 1 worked, but Fan 2 never came on. The motorist noticed no change in performance, but the vehicle would have failed the emissions test.

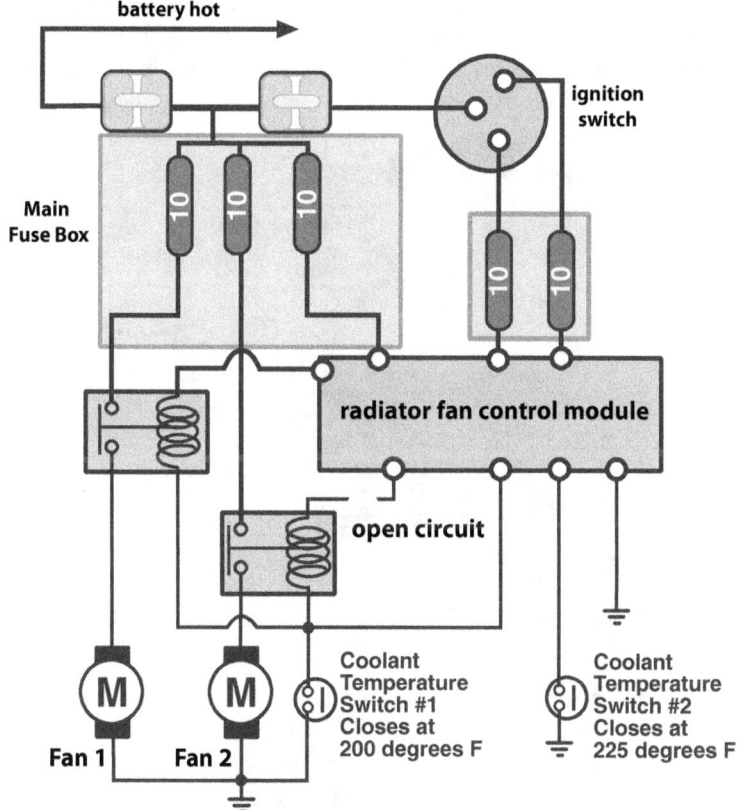

We normally think of DTCs caused by bad injectors, spark plugs, and oxygen sensors. But a cooling fan DTC will turn on the MIL and fail a car just as quickly.

# NOx

## Cooling Systems and NOx

• **Make sure the correct thermostat is installed and that it is working properly.** Thermostats that stick closed or partially closed restrict coolant flow. This increases coolant temperature, raising NOx levels.

• **Check pressure caps.** Pressure test them, don't guess. Even if the coolant overflow has plenty of coolant in it, the radiator level may be low if the pressure cap lower seal is leaking. Replace damaged caps only with the correct pressure-rated cap.

• **Check cooling fan condition and operation.** Replace damaged fans. Beware of aftermarket viscous fans that move less air and/or spin at lower rpm at certain temperatures. Check the operation of electric cooling fans. Make sure they run on both low and high speeds where applicable.

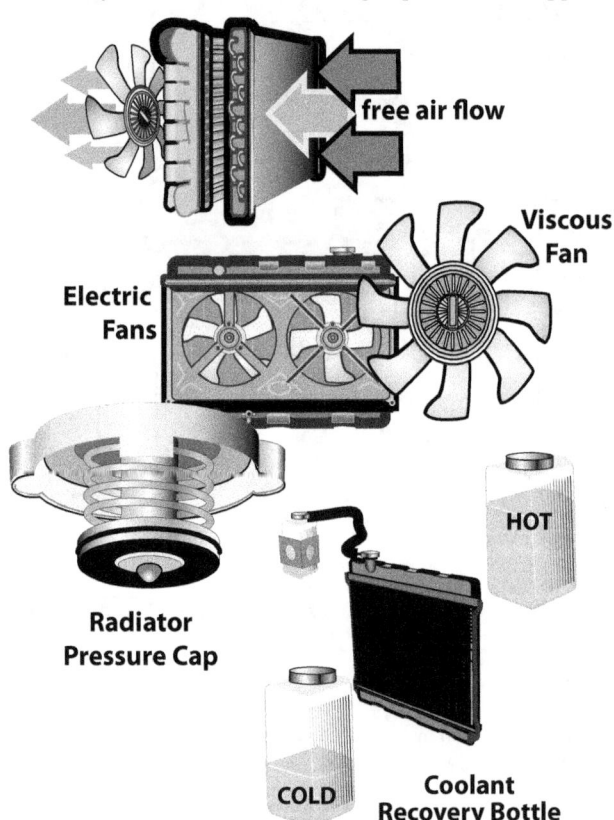

## Cooling Systems and NOx

By now, it should be very clear that cooling system efficiency has a HUGE impact on NOx.

Here's a cooling system checklist for NOx failures:

- **Bleed ALL of the air from the cooling system.** Trapped air pockets cause localized overheating, elevating NOx.

- **Use a 50/50 mix; half water and half antifreeze.** Higher antifreeze concentrations damage water pump seals and other cooling system components. High antifreeze concentrations reduce heat transfer and may settle out as a gel, blocking coolant flow.

- **The maximum antifreeze-to-water ratio is 60/40**: 60 percent antifreeze to 40 percent (distilled or demineralized) water. If you live where that mix won't provide adequate freeze protection, you'll need a block heater.

OEMs have spent millions of dollars developing antifreeze/coolants for their engines. Ford went so far as to design new engines that would be compatible with the "fleet" antifreeze. Each OEM coolant is mixed to promote good heat transfer and protect cooling systems from corrosion. We use the OEM-recommended coolant on all repairs, and are highly suspicious of any product claiming to be a *universal* coolant.

# NOx

## Testing EGR Operation

Modern vehicles monitor EGR performance using EGR valve position sensors, temperature sensors, or even Differential Pressure Feedback Exhaust (DPFE) sensors. Some manufacturers open the EGR during high-speed closed-throttle deceleration and look for a change in engine MAP to check for restricted EGR flow. (See page 172.)

Here are a few tips for testing EGR operation:

- Test EGR flow at about 1800-2500 rpm. Apply vacuum directly to the EGR valve. As the valve opens, there should be a noticeable change in the way the engine runs.
- If the engine doesn't respond as expected, look for a damaged or sticking EGR valve or a blocked exhaust gas passage between the valve and intake manifold.
- When testing positive pressure EGR valves, it may be necessary to partially block the tailpipe with a rag to raise backpressure enough to allow the EGR valve to hold vacuum.

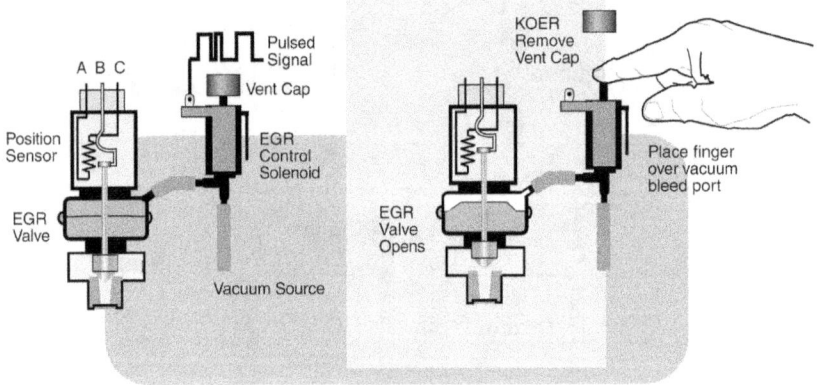

Vacuum-operated EGR systems like the Ford sonic EGR can be activated easily for testing. Locate the EGR control solenoid. Remove the solenoid vent cap and place your finger over the vacuum vent nipple. This closes the vacuum leak to atmosphere, applying vacuum to the EGR valve.

 **P0400-P0409** - EGR control, flow, or sensing fault

## Testing EGR Operation

Once you've determined that EGR flows enough exhaust gas when operated manually, we need to determine if the EGR is getting the signal to open. On vacuum-operated EGR valves:

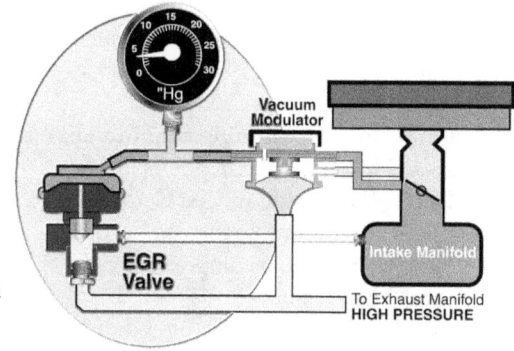

- Tee a vacuum gauge into the hose supplying vacuum to the EGR valve.
- Run the gauge into the passenger compartment or secure it to the windshield and drive the vehicle.
- You should see a vacuum signal of 2 inHg or more on moderate acceleration with the engine fully warm, at engine speeds between 1500-3000 rpm and vehicle speeds of 15 mph or more.

Use a scan tool to check all electric (linear) EGR valves. Use the scanner's bidirectional controls to operate the EGR and then test with a DMM, scope, or high impedance test light to verify that the signal from the PCM is reaching the EGR valve. (For more about linear EGR, see page 129.)

Some scan interfaces display both the command sent by the PCM and the EGR valve's response. To verify EGR flow, look for an increase in manifold pressure in datastream as the EGR opens. The graph shows how manifold pressure increases as the EGR valve is opened.

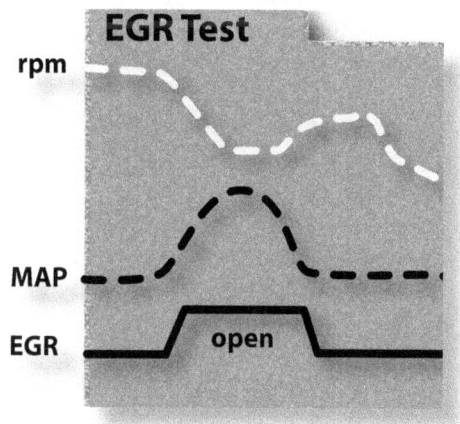

# NOx

## Common NOx Problem Areas

HELP! I fixed the CO problem but now the vehicle fails for NOx!

Relax, you're not alone: this is a frequent scenario. Here's why:

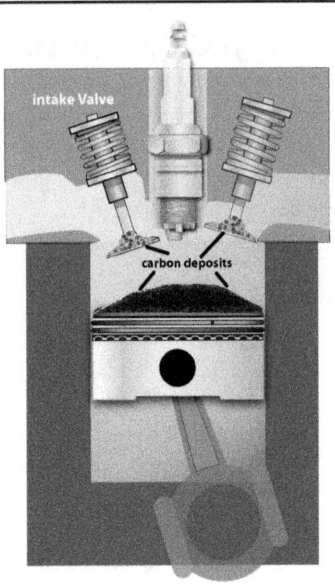

- **A previous rich-running condition has created heavy carbon deposits inside the combustion chambers.** Carbon reduces the volume of the combustion chamber, increasing compression pressure. More pressure equals more heat! More heat = more NOx. A good carbon cleaning treatment will fix things up in a jiffy. You may want to include carbon removal in any cost estimate for repairing a CO or NOx failure.

- **Sticking pre-heat door motors or heat-risers can also cause NOx failures.** Most TBI (throttle body injected) vehicles and some PFI (port fuel injected) vehicles use a pre-heat stove and duct system that warms the incoming air to reduce fuel condensation during cold running. If the temperature-controlled air door (vacuum- or wax pellet-operated) sticks in the HOT position, high intake air temperatures may cause a NOx failure in warm weather. Always test these doors with a heat gun or a hot/cold air blower.

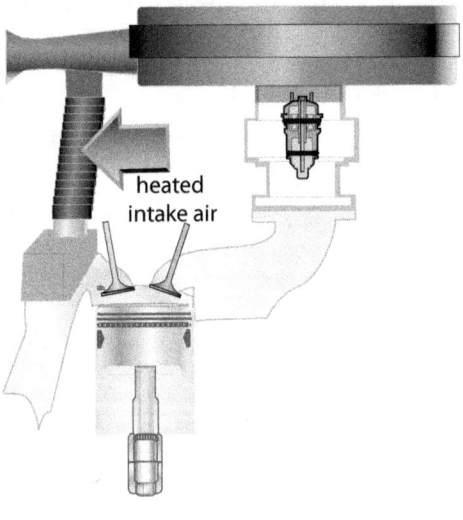

## Common NOx Problem Areas

- **Oversized tires and NOx:** Oversize tires, common on many SUVs and pickups, can cause NOx levels to skyrocket. Tires that are more than one size larger than the originals can add enough rolling resistance to overload the engine and cooling system, increasing heat.

- **The wrong spark plug heat range can drastically alter emissions.** Possible symptoms include: plug fouling or melting (yikes!), high HC emissions, misfire, or high NOx levels. For best results, use the original equipment plug (but check for TSB updates!).

- In general use, a spark plug with the OEM-recommended tip configuration and plug heat range is your best choice.

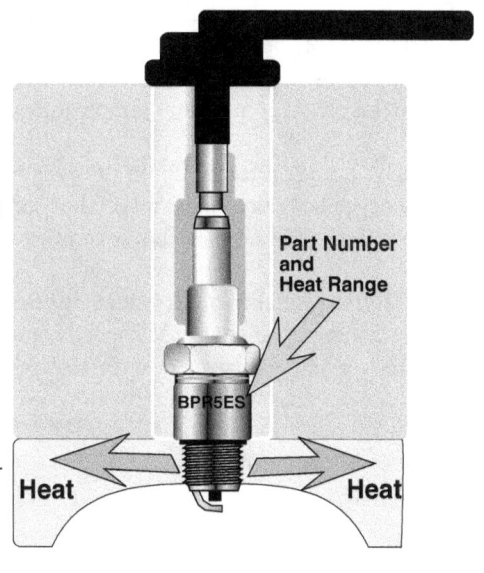

Part Number and Heat Range

BPR5ES

Heat          Heat

# NOx

## Miscellaneous NOx Tips

- **If there is a severe NOx failure, don't rush to blame the catalyst.** Reduction catalysts are improving all the time, but the place to control NOx is ahead of the catalyst, by limiting combustion chamber temperatures.

- **Clean those EGR passages.** Honda 2.2L four cylinder engines will fill the intake EGR passages with carbon, and must be cleaned. Also look out for Fords with plastic intake manifolds (5.4, 4.6, 4.2 and 3.8 Windstar engines). The fix? Remove the intakes and clean them.

- **Replace the Chrysler backpressure EGR as a complete unit.** The EGR valve/solenoid/modulator assembly is calibrated as a unit. Replacing only part of the assembly may result in poor EGR modulation, causing engine performance problems or high NOx.

- **Fill the tank with high test!** High octane fuel burns more slowly than low octane fuel. This reduces combustion pressures and temperatures. On engines with adjustable ignition timing, make sure the timing is not over-advanced.

- **Don't forget top engine cleaning.** This is becoming so important that it should probably be a preventive maintenance procedure on most vehicles. Carbon deposits accumulate inside the combustion chambers, raising compression and combustion temperatures.

- **Fix valve guide oil leaks.** Oil in the cylinders accelerates the formation of heavy deposits. It also coats the catalyst substrate, reducing catalyst action and eventually killing the catalyst completely.

- **Be selective when buying replacement catalysts.** Use only replacement catalysts that guarantee OEM capacity and performance. When in doubt, step up to a better catalyst. Some aftermarket catalyst manufacturers offer replacement catalysts with a reduction bed for vehicles that were originally fitted with oxidation-only catalysts. The difference in price is usually very small, and these retrofit reduction units add a measure of NOx safety in the emission test lane.

## Where NOx Originates

Carbon deposits from oil leaks at valve guides or incorrect PCV valve

Over- advanced ignition timing

Wrong heat range spark plugs

Lean air/fuel ratio

Wrong thermostat

Machined cylinder heads (higher compression)

Clogged radiator coolant tube and/or air flows

Restricted coolant flow

Loose fan belt and/or faulty water pump

Over concentration of antifreeze in coolant

Modifications to turbocharger controls that raise combustion pressure and temperature

Overuse of cooling system stop-leak additives

Poor lubrication: high friction: heat

Dragging Brakes

Oversized tires

Lean running makes NOx worse, since lean cylinders run hotter. A slightly rich mixture runs cooler.

**However:** Prolonged rich running creates hardened deposits inside the combustion chamber. Deposits "take up space," reducing the size of the combustion chamber. This increases cylinder pressure and heat, both of which contribute to NOx formation.

# NOx

## Older Cars

Older vehicles have special needs. Because they are older, many need a lot of long overdue maintenance, all at once. These vehicles may also have resale values less than the cost of an emissions waiver, so discretion is called for.

Here are tips and hints for dealing with automotive graybeards:

- **Check the oil level and condition.** It's amazing what an oil change will do for some engines. Check coolant level and condition and all filters, belts, and hoses.

- **If the vehicle has an automatic transmission, make sure it is in the correct gear under various loads.** Vehicles with improperly adjusted throttle valve cables may upshift too soon. Driving in a higher gear at a slower speed lugs the engine, increasing load and cylinder heat.

- **Inflate the tires properly.** Why make that old geezer work so hard on the dyno?

- **Check and clean the EGR and all EGR passages** as outlined earlier in this section.

- **Check knock sensor operation.** Remember: kill knock, kill NOx.

- **Check ignition timing.** Sometimes retarding ignition base timing by a couple of degrees is enough to reduce NOx far enough to pass the emissions test.

- **If you are dealing with a carbureted vehicle, try richening the mixture a little.** The catalyst does a better job of oxidizing CO and HC than it does of reducing NOx. Try to split the difference. Raising CO and HC slightly may also reduce NOx.

- **Fill the tank with high octane fuel.**

## General Tips

Can oil viscosity cause a driveability or emissions problem? You betcha! Many hydraulic valve lifter oil passages are small, some smaller than the diameter of a hair. Thick oil prevents the lifter from compressing as far as it should, holding valves open. This can cause misfire or even a no-start condition. Some Variable Valve Timing circuits will stick or fail to return to their rest position, setting DTCs.

> If you just had your oil changed and the MIL pops on, check the oil (and filter) used. No sense failing an emissions test because of an oil change, and trust us, it happens.

Many OEMs have started recommending high quality, low weight oils. Ford recommends 5w20 in many engines, others are using 0w20. That's zero w20, folks. Start pouring 10w40 into these engines and you'll have serious problems that may include severe engine damage.

- **Perform routine maintenance.** Cooling system maintenance is more important than ever. Periodic cleaning of MAF sensors using approved methods can eliminate many performance and emissions problems.

- **Throw out the trash!** While we seldom perform a "tune up" in the traditional sense of the term, it's becoming more important to periodically clean fuel injectors and flush away carbon deposits from the engine intake air path, including the throttle body/throttle plate, intake plenum, intake runners, idle air control circuit, and the backside of the intake valves themselves. In addition to causing performance and emissions problems, these deposits rob the engine of power and reduce fuel economy. Just be careful to observe operating instructions for your decarbonizing equipment as well as any OEM recommendations or cautions that might damage sensors or actuators in the air path.

On the following four pages, we are happy to share some tips about injector and intake path cleaning from Jim Linder, the Injector Guru.

# NOx

## Tips from the Injector GURU

Jim Linder is commonly known as the Injector Guru. Jim, who runs a respected fuel injector sales and service in Indianapolis, Indiana, has been messing around with injectors longer than he cares to admit. With his kind permission, we have borrowed his recommendations regarding correct fuel system and injector service procedures.

All injectors eventually suffer the ill effects of olefin wax, dirt, water, and fuel additives. Fuel injector system service is different from simply cleaning injectors, however. Performing a thorough fuel injection system service cures many problems, and restores peak engine performance.

For our purposes here, it pays to remember that fuel system contamination can store OBD II diagnostic codes for fuel-related misfires, sticking idle control devices, and excessive fuel corrections, among others.

OBD II isn't designed only to locate broken components. It's designed to identify potential emissions problems, and a major cause of high emissions is poor maintenance.

Here are the steps in a thorough fuel system service:

• Test fuel pressure (and volume).

• Test the fuel pressure regulator for operation and leakage.

• Flush the entire injector fuel rail, including the injector inlet screens and pressure regulator.

• Clean the injectors.

• Clean the throttle plate and idle air control valve and passages.

• Reset minimum idle air if necessary.

• Perform a PCM Relearn.

• **Change the oil to remove any fuel and liquid cleaner that have washed past the rings into the crankcase!**

252

## Tips from the Injector GURU

Two line, flow-through cleaning machines do a better job of removing system deposits than single line cleaning systems. Fuel containing a cleaning solution is circulated through the fuel rail. Some of the loosened deposits return to the flushing machine through the return line. (The vehicle fuel supply is disabled or bypassed during the cleaning process.)

Single line units or pressure can flushers that screw into the fuel rail must remove inlet screen deposits by pushing them through the injector inlet screens. That forces all loosened deposits into the injectors. If some of the deposits get caught in the injector and restrict flow, the PCM increases injector pulse width, shortening injector life.

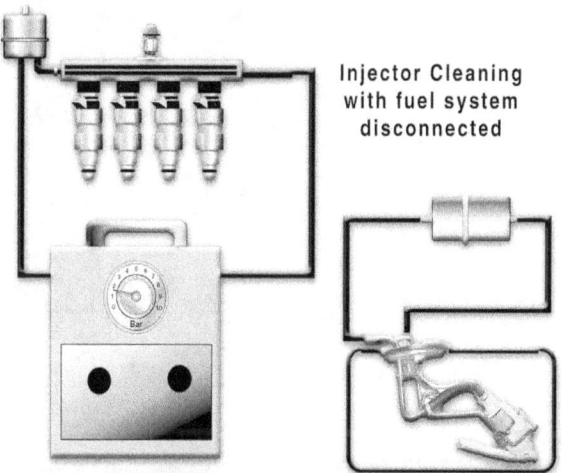

Injector Cleaning
with fuel system
disconnected

Returnless fuel systems are all the rage right now. They have no return line to the fuel tank and therefore have no throughput that can be used to flush them.

Returnless systems deadhead at the end of the line, and this can cause debris to accumulate inside the rail and at injector strainers. In extreme cases, the only cleaning option is to remove and disassemble the rail and injectors and clean them individually.

See page 187 for an image of a fuel supply system.

# NOx

## Tips from the Injector GURU

### Injector Cleaning Steps

### Step One
**Flush the rail and regulator with the engine off.** Connect the cleaning machine per manufacturer's instructions. Use the cleaning machine pump to circulate the recommended fuel/cleaning fluid solution through the fuel rail with the **engine off**. Adjust the machine delivery pressure **higher** than the fuel pressure regulator's normal pressure setting. This ensures a good flow of cleaning chemical through the regulator and helps to soften and wash away dirt and deposits from the injector inlet screens.

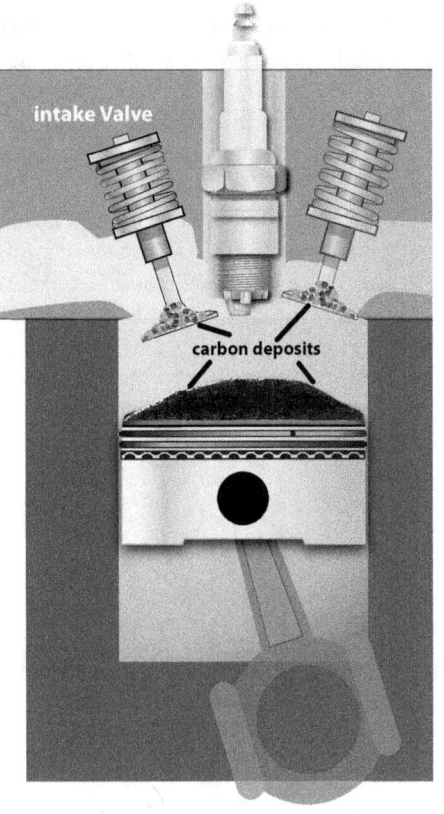

> Let the cleaning machine run long enough to do a thorough job of dissolving and removing deposits. The engine-off cleaning works best on a warm engine. Cleaning times vary from 15-60 minutes, depending on vehicle make and model, mileage, and your own experience with the quality of the local fuel supply!

### Step Two:
**Clean the injectors.** Start the engine and readjust the cleaning machine pressure slightly **lower** than the vehicle's normal, regulated pressure. Lowering the pressure slightly causes the PCM to hold the injectors open longer during the cleaning process.

> Cleaning the injectors in the engine also helps to remove carbon deposits from intake valves, the tops of pistons, and oxygen sensors. This improves air flow and air/fuel mixing for improved combustion efficiency.

The Auto Emissions Bible

## Tips from the Injector GURU

### Step Three:

Clean the air passages. The PCV, throttle plate and idle air control valve/ passages should be included in the cleaning process. Accumulated deposits in these passages strangle the engine. (We just improved fuel delivery by cleaning the fuel rail and injectors, why not make sure we have enough air to go with it?) Spray these areas with the cleaning solution, or allow a regulated amount of the solution to be sucked through the throttle in a running engine. This cleans the throttle plate, IAC, and crankcase ventilation system as the engine runs.

> The improvement in engine breathing is usually an eye opener in itself. IAC counts that had previously hovered in the 40+ range commonly drop to 15. That's why you'll want to double check the minimum idle and adjust it if necessary, followed by the appropriate idle relearn procedure for the vehicle.

Visit **Linder Technical Services** at *www.lindertech.com*

---

**Heavy Breathing**

An increasing number of new engines use dual intake manifold runners controlled by butterfly valves that can jam and stick from intake manifold deposits.

Problems include hard starting, poor performance, increased oil consumption and, of course, DTCs and the illumination of the "check money" light.

Generic DTCs associated with intake manifold runner controls (IMRC):

**P2004 to P2023** - assorted IMRC codes for control, position sensing, and performance.

---

# NOx

## PCV Valves

Misapplied PCV valves are among the most frequently overlooked contributors to driveability and emissions problems, including NOx. Honest. Symptoms range from hesitation and stalling, to uncontrolled high idle, elevated HC levels, and high oil consumption.

What does a PCV valve have to do with NOx? A faulty PCV valve can increase oil consumption if it sucks crankcase oil into the intake. This accelerates carbon deposit formation, which leads to elevated NOx levels. (Lube oil in the exhaust also contaminates catalytic converters.)

Legitimate PCV valve manufacturers list hundreds of different part numbers to cover the constellation of years, makes, and models. Beware the generic, or even "20 sizes fits all" replacement, PCV valve. Each PCV valve is carefully calibrated to the engine. PCV pellet size and weight, spring tension and inlet and outlet port configurations all affect valve performance.

DTCs commonly associated with NOx failures:
**P0325-P0335**
Knock sensor/circuit faults

**P0401-P0406 -**
Various EGR faults

DTCs commonly associated with PCV failures:
**P0505-P0507**
Idle control circuit faults

**P03xx**
Engine misfire

Tired of hearing us caution you about PCV valves? Then stop ignoring them! Seriously. When was the last time you spent good money to purchase a quality PCV valve as a part of preventive maintenance program? Thought so.

The Auto Emissions Bible

## Knock Sensors

We said earlier that NOx is an indicator of excessive combustion chamber temperature. The reduction bed of a catalytic converter can clean up a moderate amount of NOx, but manufacturers have developed a number of strategies to prevent NOx formation ahead of the catalyst to let them meet increasingly stringent test limits.

Many engines use knock sensors to detect the sonic vibrations that occur when "knocks" make NOx. These piezo sensors output a voltage signal used by the PCM to determine when knocking occurs. The PCM retards ignition timing, usually in steps, until the knock disappears.

Knock sensor operation is especially critical in turbocharged and supercharged engines where uncontrolled knock will eat holes in pistons!

Generic DTCs commonly associated with knock sensors:

**P0325 - P0334**
Assorted knock sensor codes

# NOx

## If You Hit the Wall

Sometimes you simply run into a brick wall on a diagnosis.

Odds are, your problem is on this list! Experience teaches that when all else fails, it's time to go back to basics.

Read through the list again, and then ask, "Of all the possible causes for my problem, which ones are usually associated with the emissions failure at hand. More importantly, have I tested them properly?" (See the test list on Page 10.)

Then start testing in a focused way to rule out as many causes as possible. It works.

- **Fuel Supply:** Check fuel pump; fuel tank, lines, hoses, filters; fuel pressure control; fuel distribution; fuel quality.

- **EVAP System:** Check charcoal canister; vent lines/hoses/purge valve diaphragms; fuel cap; purge valves and solenoids; mechanical control systems; electrical control systems.

- **Fuel Metering:** Check mechanical control system; electronic control system; injector(s); throttle body; idle mixture control; cold start system; oxygen sensor and heater; engine coolant temperature sensor; air flow sensor; intake air temperature sensor; throttle position sensor; MAP or BARO sensor; CKP sensor; CMP sensor; knock sensor; VSS

- **Idle Speed:** adjustment; idle air control valve.

- **Air Supply:** Check air filter; hot air intake system (TAC); intake mani fold/gasket; vacuum/false air leak; turbocharger; supercharger.

- **Ignition System:** Check ignition module; primary wiring; coils/secondary; spark plug wire; spark plugs; spark timing

- **Electrical/Electronic:** Check PCM; clear DTCs; actuators; wiring (open circuit, high resistance, shorted); battery; charging system.

## If You Hit the Wall (continued)

- **Emissions Systems:** Check catalytic converter (empty, melted, damaged, low efficiency, or damaged downstream air tube plumbing); EGR (passages; mechanical control systems, electronic controls; valves/actuators); Secondary AIR injection system (belts, pumps, bypass and diverter, switches and valves; mechanical and electronic control systems, reed valves, check valves, other valves and plumbing)

- **PCV Valve:** Include PCV system hoses and filters.

- **Engine Mechanical:** Check internal short block; cylinder head structure/gasket; camshaft(s); timing belt, chain or sprockets; valves (lifters, rockers, dirty/burnt/bent/leaking); oil seals; valve adjustment; other seals and gaskets.

- **Engine Exhaust:** Check manifold and gaskets; exhaust backpressure.

- **Engine Cooling**: Check fan; thermostat; radiators, coolers and caps; mechanical control systems; electronic control systems.

- **Vehicle Fluids**: Check all fluid levels—coolant; crankcase oil; fuel.

- **Transmission & Final Drive:** Check internal (mechanical/hydraulic); electronic control system; external controls (vacuum, cables and linkages); final drive ratio; tire size; excessive drag.

---

A smart mechanic we knew used to make repairs look easy. When asked his secret, he replied:

> I take time to check the basics. If I don't find my problem in about 15 minutes, I go back and check the basics again, just to be sure I didn't make a stupid mistake. Most problems are basic problems. I leave the hi-tech stuff for last.

> Too many people think that testing basics is beneath them; that every operation has to be brain surgery. Phooey. All failures are traceable to a basic problem.

---

## NOx Review

- NOx forms at high temperatures.

- The best place to control NOx is inside the combustion chamber.

- EGR reduces NOx formation by lowering combustion temperature.

- Neglected cooling systems contribute to NOx formation.

- Carbon deposits inside cylinders increase pressure and heat.

- Use the correct antifreeze/coolant and proper mix ratio with water.

- Kill knock; kill NOx.

- Vehicles cured of high HC and CO often fail a subsequent test for high NOx.

# DIESEL EMISSIONS

# 14

# Diesel Emissions

## 21st Century Diesels

 When your dad was a kid, diesels were pretty simple engines, loved for their long life, low-end torque, and great fuel economy. Most diesel problems were caused by mechanical problems or poor fuel quality. Most diesel electrical system problems could be diagnosed with a test light.

No more. Diesels are 21st Century high-tech. Mechanical/hydraulic diesel injection pumps are museum pieces. The current generation of car and light truck diesels uses computer-controlled **Common Rail** injection.

The good news: If you already understand gas engine emissions, you'll make the jump to diesels quickly. Common Rail injection is patterned after gas port injection, with one computer-controlled injector per cylinder. The big difference? Big pressures. Diesel high pressure fuel pumps ram fuel to the injectors though a common fuel rail. Maximum pressures now hover near 30,000 psi, in the latest generation of common rail systems. (Workers remove paint from bridge overpasses with less pressure than that!)

The injectors are different, too. Solenoid injectors are out; piezo-electric injectors are in. Piezos demand lots of watts, with some injectors operating at 150-250 volts, at amp levels that can level you if you aren't careful. This is serious business: Wear professional safety gear, use approved voltmeters, and read the manufacturer's safety warnings. Make sure you wear high voltage protective gloves, and put the old test light away before you hurt yourself.

Our purpose here will be to give you an overview of modern diesel components and emission strategies. OEM-specific engine and emissions strategies each demand a (long) book to cover exact features and specifications. But they're out there, and the numbers are growing. Look for "clean" diesels in both light and medium duty trucks, and in passenger cars with German accents.

The Auto Emissions Bible

# Diesel Emissions

## Diesels-Old Smog, New Tricks

Diesels have posed a tough emissions challenge. Their CO and HC emissions are comparatively lower than those of gas engines, but they emit a lot of soot (aka *particulate matter*, or PM), oxides of nitrogen, and sulfur. None of this is healthy for carbon based life forms, creating smog and causing respiratory illness and distress.

**Particulate Emissions (PM)** may be new to you if you don't work on diesels. Never a serious issue with gas engine emissions, tiny soot particles in diesel exhaust gas are too small to see with the naked eye, but work their way deep into our lungs, where they stay. Their accumulation is especially serious for people with respiratory problems. To paraphrase an old saying, "Out of sight, out of breath."

Capturing PM is the task of the **Diesel Particulate Filter,** a device not used in gasoline engine vehicles.

Technological improvements to the passenger car/light truck diesels are extensive and effective to the point where some diesel models are 50-state certified.

This also means that newer diesels are truly OBD II compatible. Modern diesels have a DLC, a bank of onboard monitors, and a fully functional Malfunction Indicator Light.

Over the next few pages, we'll talk about diesel engine strategies and major features, and then add exhaust treatment components. You may be surprised at how similar gasoline and diesels strategies have become.

# Diesel Emissions

## More Parts - More Systems - More DTCs

The explosion of diesel technology has added dozens of new components to the diesel landscape, many of which were common in gas engines for years. A few, like the addition of a computer-controlled throttle valve, will shock old timers.

Unfortunately for the diagnostician, all those new parts and systems have generated hundreds of diesel DTCs, both generic and manufacturer-specific. GM estimates that changes in the 2011 Duramax diesel add over 160 *new* DTCs. Before we try to give you a brief overview of sample codes, we'd probably better start by looking at how diesel hardware has changed; in engine and fuel components and their management, and in new, complex exhaust aftertreatment strategies.

### Modern Diesel Components

Some components are carry-overs from older diesels; others are new.

**Glow Plugs** - These electrically-operated resistance heaters, one per cylinder, heat intake air to improve cold engine combustion efficiency.

> The PCM or a separate controller regulates glow plug on-time. DTCs P0671 through P0678 indicate individual glow plug failures.

- **MAF Sensors** - Long used in gasoline engine fuel management, the Mass Air Flow sensor does exactly what its name suggests: it measures the mass of the air entering the engine. The PCM in a diesel needs information about air mass and oxygen available for combustion before it can calculate fuel delivery. This is no different from how a MAF works in a gas engine vehicle.

- **MAP sensors** (sometimes called *boost* sensors in diesels) are a carry-over. They measure manifold pressure/boost.

MAP signals and the vane position sensor signal in variable displacement turbocharger systems regulate and limit boost.

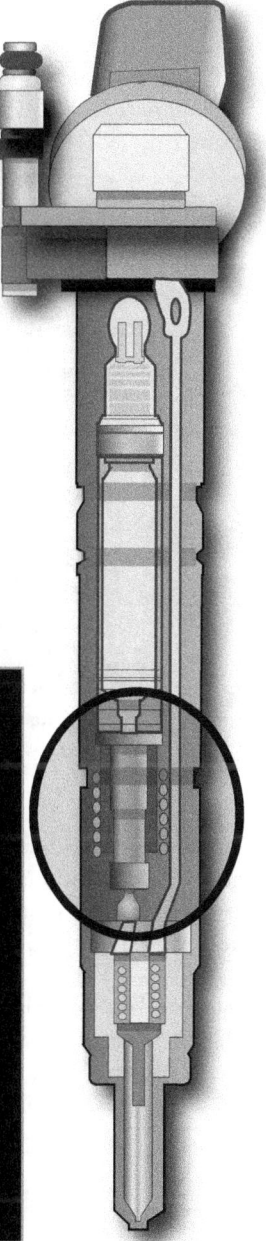

**NEW!** **Fuel Injectors** - You'll find different injector designs in Common Rail systems. Early Common Rail diesels used solenoid injectors, similar to the ones used in fuel injected gas engines. Current generation common rail injection uses **piezo** injectors. Stacks of piezo crystals inside the injector body flex slightly when electrified. This slight flexing is enough to open and close the injectors with blazing speed and split-second precision.

Piezos are so fast, that instead of delivering all of the fuel to the cylinder in a single squirt, they can divide total fuel delivery volume into *multiple mini-sprays* during a single combustion event, a process called ***pilot injection***. Multiple spray holes at each injector tip, combined with pulsed pilot injection and super high fuel pressure, mix fuel and air better, burn fuel more efficiently, increase power, and reduce noise and harmful emissions.

**Caution:** Modern injectors may be flow tested and matched into groups by performance. This coded information will be printed on a label or etched into the injector body. Injector flow characteristics are then programmed into control units to optimize engine performance.

Check OEM recommendations when replacing injectors. You may be required to use a scan tool and enter the new injector identification numbers into the PCM.

A similar situation may exist with glow plugs and their control modules. Parts swapping and failure to properly reprogram a controller when new parts are installed may lead to DTCs and poor engine performance. We really need to discard some old thinking when we start diagnosing modern diesel concerns.

# Diesel Emissions

**EGR** - Diesel EGR works like gas engine EGR: it introduces a measured amount of exhaust gas back to the intake manifold to limit peak combustion temperature. Lower temperatures reduce NOx emissions. To improve EGR efficiency, diesels have large liquid-cooled EGR systems that remove heat energy from exhaust gases before they are introduced to the intake air stream.

This cutaway shows how the EGR coolers are encased in a water jacket. - image Crooked River Writer

**Temperature Sensors** - Again: this is very similar to gas engine design. Diesels have the normal array of sensors reporting the temperature of various fluids to the engine computer. Air, fuel, engine oil and coolant temperatures are all monitored, as are the sensor circuits.

Liquid coolant works hard in diesels. In addition to cooling the engine, it pulls excess heat from the charge air cooler, liquid-cooled EGR valves, engine oil coolers, and turbochargers. Expect to see dual cooling systems and dual range thermostatic control of engine coolant levels in different zones.

 Example: The Ford 6.7L has two thermostats with staggered opening temperatures placed side by side in a single thermostat housing. The primary coolant circuit pulls heat from the engine, EGR and oil coolers, and the turbocharger. The secondary circuit circulates coolant through the EGR, transmission, engine oil, and charge air coolers. This system even has primary and secondary water pumps.

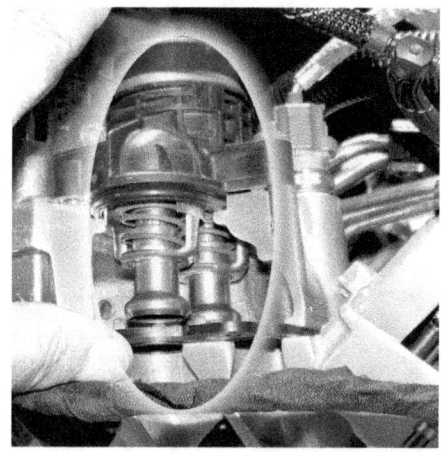

Ford PowerStroke twin thermostats. - image Crooked River Writer

## Throttle Valve

A throttle on a diesel? What is this world coming to?

This throttle restriction valve closes under special circumstances: during cold-start enrichment and during Diesel Particulate Filter (DPF) regeneration.

Ford PowerStroke throttle valve. - image Crooked River Writer

Closing the valve also increases manifold vacuum (decreases manifold pressure) to improve EGR flow.

# Diesel Emissions

## Diesel Monitors

The monitor list for diesels is a little shorter than it is for gas engines, but when trouble starts, we can still round up some of our favorite suspects.

- **Comprehensive Components** - The powertrain control module continuously tests the computer, plus sensor inputs and actuator outputs. All must operate within test limits, and make sense. Commands are compared to responses to ensure that they are logical.
- **Misfire** - As it is with its gas engine counterpart, the misfire monitor uses the crankshaft sensor speed input to look for drops in crankshaft rpm caused by misfire.
- **Fuel System** - All fuel sensors and actuators are monitored for shorts and opens. Other commonly monitored conditions include: injector supply voltage and current, and fuel temperature and pressure.
- **Cooling System** - With some cooling systems now operating with dual thermostats, water pumps, and cooling zones, expect to see multiple engine coolant temperature sensors.
- **EGR Monitors** - EGR and diesels go together like chocolate and peanut butter. Diesel compression ratios are very high and create NOx quickly. That's why diesels use EGR coolers to lower exhaust gas temperature for improved NOx control.

## Monitored Components

Now let's look at components commonly tested by OBD diesel monitors. Gas engine guys will find this largely familiar. OBD II has monitored these components and their circuits for years in gasoline engines.

- Fuel Injector Control Circuits
- Fuel Injector Voltage Control Circuits
- Accelerator Position Sensor
- Throttle Position Sensor
- Crankshaft Angle Sensor
- Camshaft Angle Sensor
- Manifold Absolute Pressure (MAP) Sensor
- Barometric Pressure (BARO) Sensor
- Intake Air Temperature Sensors
- Fuel Temperature Sensor
- Engine Coolant Temperature Sensor
- Mass Air Flow (MAF) sensor
- Turbocharger Vane Position Sensor
- Turbocharger Vane Control Solenoid
- Glow Plug Control Module and Glow Plugs
- Memory (ROM/RAM)
- Malfunction Indicator Lamp (MIL) Monitor
- Bus or Local Network Communications
- Vehicle Speed Sensor (VSS)
- NOx Sensor
- $O_2$ sensor (example: the VW Diesel Oxidation Catalyst has an oxygen sensor)

# Diesel Emissions

## Diesel Fuel and Emissions

Traditional middle distillate petroleum diesel fuel chemistry has long contributed to harmful diesel emissions. Even Low Sulfur Diesel has enough sulfur at 500 ppm to quickly contaminate diesel exhaust scrubbing devices like catalysts and exhaust filters. Before we could get serious about cleaning diesel exhaust, we had to clean the fuel.

Beginning January 1, 2007, the 2007 Highway Diesel Rule mandates a 50% reduction in NOx and HC, and a 90% reduction in particulate matter (PM), below 2004 levels. The standards were cinched tighter in 2010. To make this all possible, diesel fuel underwent a really radical change to reduce its sulfur content by a whopping 97 percent—from 500 ppm—to only 15 ppm!

Even without changes in diesel engine design or the addition of special emission control devices, **Ultra Low Sulfur Diesel (ULSD)** fuel reduces harmful emissions immediately; more importantly, it no longer contaminates diesel emission scrubbing devices and filters.

Fuel dispensers identify **Ultra Low Sulfur Diesel (ULSD)** with the label shown here. Use of diesel fuels with higher sulfur content will damage diesel emission components. Component damage may include catalyst poisoning and soot clogging of the large exhaust filter referred to as the **Diesel Particulate Filter (DPF)**.

Crankcase oil has also been changed to lower its ash content. Use the lube oil recommended by the vehicle maker. Oil grades with a higher ash content clog the DPF much faster.

OEMs are increasingly embracing biofuels. Many OEMs originally limited biofuel use to **B5** (5% biofuel added to 95% petroleum-based fuel). Some OEMs are now endorsing the use of **B20** (20% biofuel added to 80% petroleum-based fuel).

# Diesel Emissions

## Diesel Exhaust Cleaning Components

**NEW!** This overview shows a common diesel exhaust system layout. Far more complicated than those found on pre-emissions diesel exhausts, the modern diesel exhaust system contains several unique components, each with its own specialized task.

Diesel vehicles may use some or all of the components illustrated below and explained on the facing page, depending on year, make, and model.

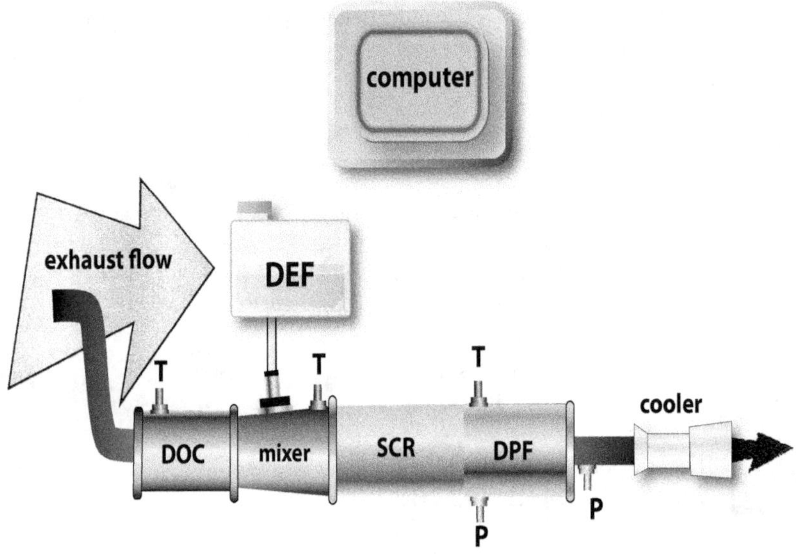

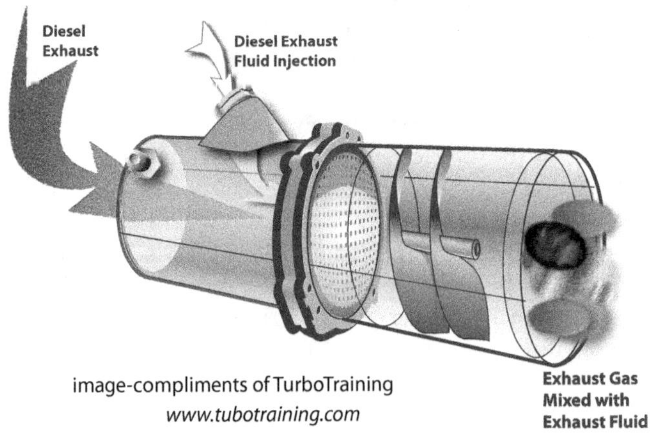

computer

exhaust flow

DEF

cooler

T     T     T

DOC  |  mixer  |  SCR  |  DPF

P     P

Diesel Exhaust

Diesel Exhaust Fluid Injection

image-compliments of TurboTraining
*www.tubotraining.com*

Exhaust Gas Mixed with Exhaust Fluid

 ## Diesel Exhaust Cleaning Components

- **Computer** - The engine management computer monitors exhaust temperature and pressure at several points. It also determines when the DEF injector adds urea to the exhaust system to reduce NOx emissions.

- **Diesel Oxidation Catalyst** - The DOC works like the catalyst found in gasoline powered vehicles. It oxidizes HC and CO to produce water and $CO_2$.

- **Diesel Exhaust Fluid** (aka Reductant Fluid) - On command from the PCM, Diesel Exhaust Fluid (DEF) is injected into the exhaust to reduce NOx emissions. The tank must be filled periodically with a special DEF fluid. A low fluid level may trigger a limited operating strategy that reduces engine speed until the fluid tank is replenished.

- **DEF Mixer** - The exhaust may be fitted with a mixer that swirls the exhaust and the diesel exhaust fluid, promoting better reduction.

- **Temperature Sensors** - Multiple temperature sensors (T) report back to the engine computer.

- **Selective Catalyst Reduction** - SCR reduces $NO_X$, creating harmless nitrogen.

- **Diesel Particulate Filter** - The DPF collects and incinerates soot (particulate matter). The cleaning process is called **regeneration**, a process that burns away accumulated soot; it works a lot like a self-cleaning oven. When sensors indicate that the filter is full, a special incineration process burns away the contents. DPF cleaning occurs automatically (passive cleaning) or on command from the computer (active). A separate service mode cleaning process may be activated by scan tool command.

- **Pressure Sensors (P)** - Sensors at the DPF inlet and outlet measure the pressure drop across the DPF to determine when regeneration is needed.

- **Tailpipe Cooler** - The exhaust can get VERY hot, especially during regeneration. Tailpipes are commonly fitted with a venturi cooler.

# Diesel Emissions

## Sample Diesel DTCs

We cannot list all of the DTCs that you might see associated with the new clean diesels, but we wanted to give you a sampling of generic diesel codes. That way you can see the similarities between familiar gas engine codes and those used on diesels.

**P0087** - **Fuel Rail Pressure Too Low** - Detects low fuel rail pressure.

**P0088** - **Fuel Rail Pressure Too High** - Detects high fuel pressure.

**P0090** - **Fuel Pressure Regulator Control Circuit** - Detects opens and shorts in the Fuel Pressure Regulator Control Circuit.

**P0106** - **MAP Sensor Range and Performance** - This code indicates that the MAP exceeds test standards or that MAP and BARO sensor values do not agree KOEO or KOER.

**P0107** - **Manifold Absolute Pressure (MAP) Sensor Circuit Low** -Detects low voltage in the pressure sensor circuit, commonly caused by a short.

**P0108** - **Manifold Absolute Pressure (MAP) Sensor Circuit High** - Detects low voltage in the pressure sensor circuit, commonly caused by an open.

**P0128** - **Coolant Temperature Below Thermostat Operating Temperature** - This indicates a cold running engine, commonly caused by a coolant thermostat problem.

**P0191** - **Fuel Rail Pressure Sensor Performance** - Compares fuel rail pressure to a reference pressure (commonly atmospheric pressure) to detect a range or performance problem in the fuel rail sensor.

**P0192** - **Fuel Rail Pressure Sensor Circuit Low Voltage** - Detects low voltage in the pressure sensor circuit, commonly caused by a short.

**P0193** - **Fuel Rail Pressure Sensor Circuit High Voltage** - Detects high voltage in the pressure sensor circuit, commonly caused by an open.

**P0403** - **Exhaust Gas Recirculation Control Circuit** - Test for shorts and opens in the EGR control circuit; includes components and wiring.

**P0405** - **Exhaust Gas Recirculation Position Sensor Circuit Low Voltage** - Tests the EGR position sensor circuit that provides feedback about EGR valve true position.

**P0406** - **Exhaust Gas Recirculation Position Sensor Circuit High Voltage** - Tests the EGR position sensor that provides feedback about EGR actual position..

**P0401** - **Exhaust Gas Recirculation (EGR) Flow Insufficient** - Uses other sensors (commonly the mass air flow sensor, or MAF) to compare air intake mass to EGR volume at a given EGR valve opening. Commonly indicates **low** flow due to EGR valve restriction or clogged EGR passages.

**P0402** - **Exhaust Gas Recirculation (EGR) Flow Excessive** - Uses other sensors (commonly the mass air flow sensor (MAF) to compare air intake mass to exhaust gas volume at a given EGR valve opening. Indicates **excessive** flow due to EGR valve malfunction.

**P046C** - **Exhaust Gas Recirculation (EGR) Position Sensor Performance** - This test checks the EGR position sensor to confirm that the valve moves to its commanded position. Used to diagnose a sticking or malfunctioning EGR valve.

Ford PowerStroke common rail.-
image Crooked River Writer

## Diesel Emissions Review

- Diesels are now true OBD II vehicles.

- Diesels have a DLC and MIL, store DTCs, and run monitors.

- Many of the components that make diesels run cleaner are carried over from gasoline engine technologies.

- Ultra Low Sulfur Diesel (ULSD) fuel contains far less sulfur than Low Sulfur Diesel fuel.

- The use of ULSD fuel allows the use of catalysts and exhaust cleaning devices that incinerate diesel particulate matter.

- Common rail fuel systems with pulsed pilot injection improve combustion efficiency, run quieter, and have more power.

# INFORMATION 15

# Information

## Repair Information

Surprised to see **Information** listed first in our toolbox? Don't be. Without access to information, you have none of the following:

- DTC Definitions.
- Test standards used to pass and fail monitors
- Repair standards
- Technical Service Bulletins (TSBs)
- Computer software updates

Here are several sources of information; some of it is free, some is pay-per-view or pay-per-download.

### NASTF

The National Automotive Service Task Force NASTF) is a clearing house for OEM service and repair information. It provides links to major vehicle manufacturers' service web sites where you will find repair manuals, newsletters, and software downloads.

While some generic information is free, expect to whip out the old credit card for the good stuff, including software downloads required for reflashing the vehicle computer.

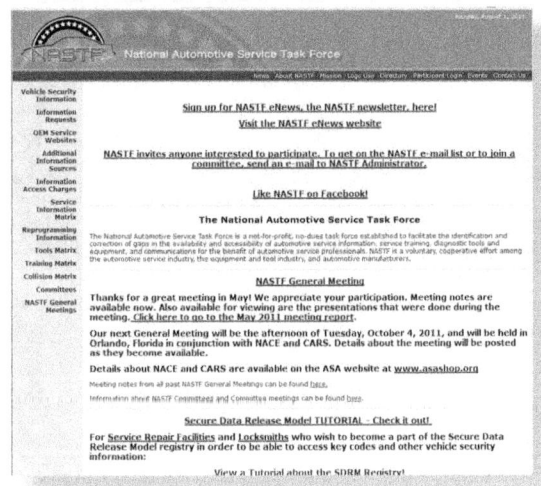

# Information

## iATN

Most professional auto repair technicians already know about the International Auto Repair Technicians' Network (iATN), but it is too valuable an asset not to be included here.

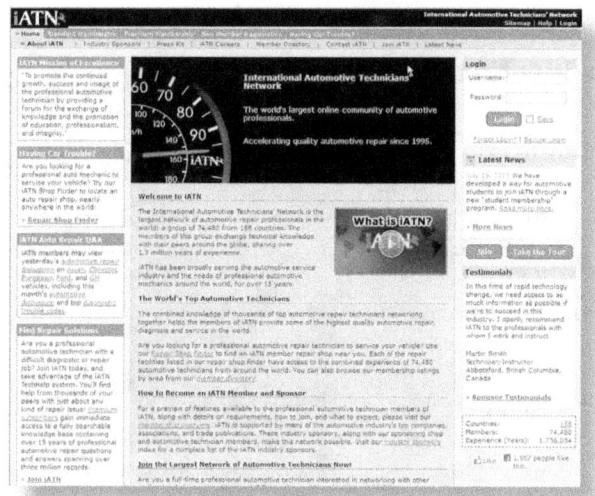

From their web site:

> The International Automotive Technicians' Network is the largest network of automotive repair professionals in the world: a group of 74,480 from 155 countries. The members of this group exchange technical knowledge with their peers around the globe, sharing over 1.7 million years of experience.
>
> iATN has been proudly serving the automotive service industry and the needs of professional automotive mechanics around the world, for over 15 years.

This amazing repository of training, raw repair information, and personal experience does a great job of matching repair symptoms to vehicle-specific repair solutions. Those who prefer systems-based repair strategies will also enjoy the give and take of forum discussions on automotive theory and practice, and a large and growing reference library.

Basic membership is still free (amazingly!), although value-added premium membership and sponsorships are available.

Membership is limited to individuals with professional automotive credentials. But even if you do not qualify as a member, odds are someone you know is a member, and can help you get answers to tough repair questions.

# Information

## Identifix

Identifix is a subscription-based auto repair information service.

www.identifix.com

Subscribers access information either through the Identifix Hotline or through the Indentifix Direct Hit online database that stores thousands of verified vehicle repairs.

From their web site:

> Since its founding in 1987, Identifix, Inc. has evolved into the nation's best source for knowledge on what breaks, which vehicles it breaks on, and how to fix those vehicles correctly. This knowledge is derived from continual analysis of data gathered from nearly 250,000 calls received each year from technicians seeking assistance in diagnosing vehicle repair problems; a staff of 45+ factory certified, ASE master technicians; and the nation's most comprehensive on-site library of factory vehicle service information.

Identifix members also have access to library reference materials such as component locators and wiring diagrams.

## Repair Databases

The best known paid subscription professional repair information services in the U.S. provide subscription access to large databases containing a combination of how-to procedural information and vehicle repair and test standards.

• Your local library may have a subscription available to library cardholders.

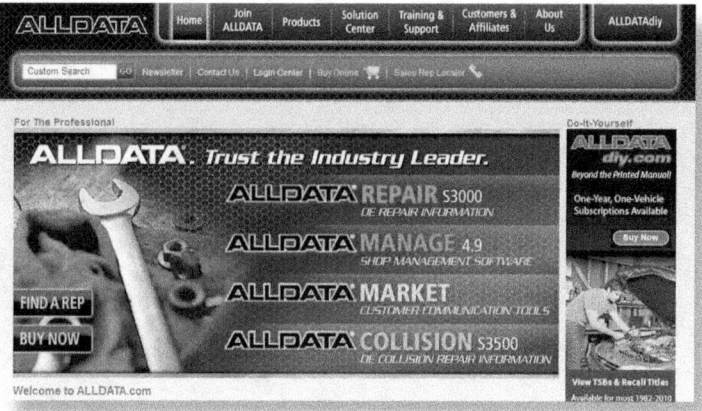

AllData offers reasonably priced one-year subscription rate on a per-vehicle basis. http://www.alldata.com/

Mitchell OnDemand DIY offers a limited subscription program, offering daily, weekly, or monthly access for a single vehicle. http://www.ondemand5.com/

## Notes

# GLOSSARY

- **Adaptive Memory** - Adaptive memory stores data about adjustments the Powertrain Control Module (PCM) makes to compensate for changes in operating conditions and component wear. Data are stored in Keep Alive Memory (KAM) and are not erased when the ignition is switched off.

- **AIR - (Air Injection Reaction) system** - An abbreviation for the secondary air management system, an air pump and valves that inject air into the exhaust manifold ahead of the catalyst to oxidize unburned fuel.

- **Air Flow Metering -** A fuel control strategy that determines the amount of fuel needed for combustion primarily by measuring the volume or the mass of air drawn into the engine.

- **Alphanumeric** - *Alphanumeric* describes the OBD II trouble code numbering system. Diagnostic Trouble Code (DTC) designations are a combination of letters of the alphabet and numbers: Alpha + numeric. OBD II DTCs are alphanumeric, as in P0300.

- **Ambient Temperature or Pressure** - The temperature or barometric pressure of the air surrounding the vehicle.

- **Bidirectional** - A bidirectional scan tool interface transmits data in both directions. In a bidirectional interface, the scan tool can receive data from the controller and also transmit data back to the controller. In a scan tool interface, the scan tool can request data, retrieve and erase Diagnostic Trouble Codes, and command the PCM to operate various vehicle components for test purposes.

- **Catalyst** - A *catalyst* is a something that promotes a chemical reaction without being consumed or changed by the chemical process. The vehicle catalytic converter uses this characteristic to promote chemical reactions that clean the exhaust, without being consumed in the process.

# Glossary

- **Closed Loop -** For our purposes here, closed loop is a control strategy where fuel delivery is continually adjusted to produce an air/fuel mixture that hovers near a stoichiometric balance. The system is referred to as "closed" because fuel adjustments are based in part on feedback from an oxygen sensor located in the exhaust stream.

- **Comprehensive Component Monitor (CCM)** - The CCM is a Continuous Monitor that tests the electrical integrity and function of inputs (sensors) and outputs (actuators).

- **Continuous Monitor** - A monitor that runs... *continuously* during normal vehicle operation, as opposed to a Non-Continuous Monitor that runs to completion and stops, once per trip.

- **DLC** - Data Link Connector. The term DLC has been used before by some manufacturers, but in OBD II, it represents the standardized 16-pin diagnostic test receptacle used to connect the OBD II scan tool to the vehicle computer.

- **DTC** - Diagnostic Trouble Code

- **ECT** - Engine Coolant Temperature sensor

- **Enabling Criteria** - The exact conditions needed to run a monitor. These pre-conditions are provided by a *trip*.

- **Enhanced EVAP** - An Enhanced Evaporative Emission System uses monitoring strategies that test fuel vapor purge flow to the engine, and check the fuel containment system for vapor leaks to the atmosphere.

- **EVAP** - The Evaporative Emission system.

- **Freeze Frame** - A single frame snapshot of critical engine operating conditions, usually stored with the first Diagnostic Trouble Code (DTC). Freeze Frame is normally stored with the first DTC, although Freeze Frame from a previous DTC may be overwritten by Freeze Frame from a later DTC if it represents a more serious problem. Vehicles must store at least one Freeze Frame, although some will store multiple Freeze Frames.

# Glossary

- **Fuel Trim -** A numeric representation of the fuel corrections needed to keep the air/fuel ratio at stoichiometry. Fuel trim may be **short-term** or **long-term**, referred to as STFT and LTFT, respectively. STFT represents corrections occurring in closed loop. LTFT generally represents long term trends in STFT. Total fuel trim = LTFT + STFT.

- **HC** - Hydrocarbons (fuel).

- **IAT** - Intake Air Temperature sensor

- **Inches of Water** - A measurement standard commonly used to measure small pressures inside the EVAP system. One PSI is equal to 27.68 inches of water.

- **KAM** - Keep Alive Memory (KAM) retains information in PCM memory even after the ignition is switched off, since it is powered by constant battery voltage.

- **KOEO** - Key-ON-Engine-OFF.

- **KOER** - Key-On-Engine Running.

- **Lambda -** Lambda is a value that indicates how close the air/fuel ratio is to 14.7:1 (stoichiometry). At lambda 1.00, the mixture is in stoichiometric balance. Numbers greater than 1 indicate a leaner mixture; those smaller than 1 indicate a richer condition. Lambda is the most accurate way to measure air/fuel ratio, since it not affected by combustion, or the lack of it.

- **limp home** - A special operating strategy designed to keep the vehicle running well enough to "limp home" in the event of a serious failure or loss of a critical sensor input. Vehicle performance is often limited in this mode.

- **LTFT** - Long Term Fuel Trim (LTFT) is a learned value stored in Keep Alive Memory (KAM) about long term fuel corrections. LTFT values reflect STFT fuel trends over time.

- **MAF** - The Mass Air Flow sensor is a device that measures the mass of air drawn into the engine. KOEO and wide-open-throttle MAF readings may also be used by the PCM to measure barometric pressure.

# Glossary

- **MAP** - Manifold Absolute Pressure is the pressure inside the intake manifold. It is the difference between atmospheric pressure (BARO) and the pressure/vacuum inside the intake manifold. MAP is also used as a shorthand reference for the MAP sensor.

- **MAP Sensor** - MAP sensors measure intake manifold pressure as an indication of engine load. MAP sensors may also be used in MAF vehicles as a diagnostic sensor and backup signal for the MAF.

- **MIL** - The Malfunction Indicator Light, also referred to as the Check Engine Light (CEL).

- **Misfire** - Interruption of the combustion process.

- **Monitor** - A test performed by the on-board diagnostic system to check the performance of various components and subsystems.

- **OBD II** - OBD II stands for On-board Diagnostics, Generation Two. This on-board diagnostic system has been standard on passenger cars and light trucks sold in the U.S. since 1996, although there were several vehicles equipped with a transitional OBD II on a selective basis, beginning in 1994. OBD II includes several test functions that detect deterioration of powertrain components or emission controls that may result in increased vehicle emission levels.

- **OEM** - Original Equipment Manufacturers; vehicle makers.

- **Oxidation** - The chemical process of adding oxygen to a compound. Fast oxidation is called burning.

- **Oxygen Sensor (HO2S Heated Oxygen Sensor)** - A voltage generator that creates a signal voltage indicating the amount of oxygen in the exhaust, compared to the amount of oxygen in the air surrounding the vehicle. Changes in oxygen sensor signal voltage inform the PCM about the air/fuel ratio. In OBD II vehicles, oxygen sensors are also installed at the catalyst outlet to measure the oxygen storage efficiency of the catalyst bed.

- **Oxygen Sensor Heaters** - Small resistance-type heaters that bring the oxygen sensor to operating temperature quickly.

# Glossary

- **Pending Code** - The first detection of a fault that uses two-trip logic. A pending code for a monitor other than fuel or misfire is retained in memory and matures into a Diagnostic Trouble Code (DTC) if the fault is detected again on the next consecutive trip. Fuel and misfire faults can mature into a DTC on non-consecutive trips.

- **PCM** - Powertrain Control Module

- **PID** - **P**arameter **ID**entification. PIDs are individual items in the scan tool data display.

- **Prosumer** - A tech savvy-consumer who has the requisite skills, information, and equipment needed to perform many repairs that usually demand the services of a professional repair technician.

- **Readiness Status** - (also referred to as Monitor Status) - A list of monitors displayed on a scan tool indicating which monitors have run to completion. Readiness Status does not indicate if a monitor passed or failed, only if it ran to completion.

- **Reduction** - The opposite of oxidation. Reduction is a chemical reaction that *removes* oxygen from a compound.

- **Serial Data** - Information transferred by a series of voltage pulses.

- **Short Term Fuel Trim (STFT)** - The amount of PCM fuel correction needed to keep the air/fuel ratio at stoichiometry in closed loop.

- **Speed Density** - A fuel control strategy based primarily on engine speed and air density.

- **Stoichiometry** - The conditions under which a particular chemical reaction proceeds most efficiently. For gasoline engines, stoichiometry occurs at an air/fuel ratio of 14.7 to 1, by weight.

- **Trip** - A key cycle (Key-ON, start to-run, and Key-OFF) that includes all driving conditions needed to run a monitor.

# Glossary

- **VSS** - Vehicle Speed Sensor

- **Warm-Up Cycle** - Warm-up cycles are used by the Powertrain Control Module (PCM) to erase Diagnostic Trouble Codes (DTCs) after the Malfunction Indicator Lamp (MIL) is extinguished by the PCM. The PCM turns off the MIL if the monitor that set the DTC runs and passes on three consecutive trips. After 40 warm-up cycles without any additional faults, the DTC is erased from memory. A warm-up cycle is defined as a start-to-run where the engine coolant temperature starts below—and then rises above—160°F, and increases by at least 40°F. No additional faults can occur during this warm-up period for the warm-up cycle to be counted.

# OBD II MODES

OBD II originally had nine primary operating Modes, defined by SAE paper J1979. A new Mode 10 for Permanent Codes has the designation $0A. The dollar sign before each mode number is normal notation, included here so you'll be familiar with it when it appears in reference materials or in your repair database.

The first four modes are special; they all appear in any properly functioning generic scan tool interface. These "core four" are the heart of OBD II diagnostic functions, and you can count on them to show up for work every time.

The other five? They are far less dependable; don't be surprised to see vehicles where Modes 5-9 are not supported, at all! Especially last century vehicles.

### Mode $01 - Request Current Powertrain Diagnostic Data
**Mode 1** displays diagnostic data from PIDs (Parameter IDentifications) that represent components and systems monitored by OBD II. Don't stumble over the term "PID"; these are nothing more than individual data items that can be viewed in datastream. PIDs display test values for sensors, actuators, monitor and MIL status, and some computer calculations and adjustments. PID data assist us when we need to find and repair problems that turn on the check engine light (MIL).

All values shown in Mode $01 display the actual status of the monitored component, never a substituted value. This is extremely important. It is one of the main reasons we need to review and study generic data as well as manufacturer-specific data.

### Mode $02 - Request Powertrain Freeze Frame Data
Mode 2 is a cross between a data logger and a surveillance camera. It records Freeze Frame "snapshots" of PID data associated with one or more DTCs.

# OBD Modes

**Example:** If an emissions-related fault is detected by the vehicle computer, the computer may store a DTC in memory. The PCM also grabs a snapshot of associated vehicle data. Click: caught in the act! Later, we can review this security data picture for clues from the scene of the crime.

Originally, OBD II vehicles stored a single Freeze Frame with the first or most severe DTC, in the event that multiple code are stored. Expect newer vehicles to store multiple Freeze Frames.

Like generic datasteam, Freeze Frame data are actual measurements, not substituted values.

### Mode $03 - Request Emission-Related Powertrain DTCs

**Mode 3** displays DTCs on an OBD II scan tool. Code numbers correspond to specific fault types. Note: The PCM may turn off the MIL; if this happens, a history code remains in memory for 40 warm-ups after the MIL is turned off.

### Mode $04 - Clear/Reset Emissions Related Diagnostic Information-

**Mode 4** clears all emissions related diagnostic data on request, and usually turns off the MIL (see Mode 10 for the exception). Emissions related data may also be cleared if power is removed from the PCM long enough to erase Keep Alive Memory (KAM).

• **Erasing emissions data:**
• Turns off the MIL.
• Clears Diagnostic Trouble Codes.
• Clears Freeze Frame data.
• Clears all test data for Modes 5, 6, and 7.
• Resets non-continuous monitors to incomplete.
• In some cases, resets fuel trim.

**Note:** The vehicle computer may not respond to a command to erase emissions data under all conditions; for example: when the engine is running. Try shutting off the engine and repeating the command to erase DTCs with the ignition key on and engine off (KOEO).

The remaining modes may or may not be available in your vehicle/scan tool interface. It depends on year, make, and model, and your scan tool's capabilities. These modes may be potentially useful for diagnostics.

### Mode $05 - Retrieve Oxygen Sensor Test Data

Mode 5 is supposed to display the results of the most recent oxygen sensor onboard tests. (The key words in the previous sentence are: "supposed to.") Don't be too surprised if the PCM shrugs its shoulders and ignores your request for this data. CAN vehicles that support this data may display it in Mode 6.

If you are lucky enough to see these values, remember that they are the most recently completed onboard oxygen sensor test results, not live oxygen sensor data. Use datastream (Mode 1) for that.

### Mode $06 - Request On-Board Monitoring Test Results for Non-Continuously Monitored Systems

In the early years of OBD II, Mode 6 was viewed as a great, dark mystery, understood only by sorcerers and wizards. Early on it was thinly and inconsistently supported by vehicle and scan tool manufacturers. If it did show up on the scan tool screen, there was little or no reference material to help you decode it. Gradually, Mode 6 has gotten more attention and support.

When it is available, Mode 6 shows the pass-fail status of onboard tests for non-continuous monitors, and some continuous monitors on select makes. It displays recent test values and test limits, helping us identify monitors that are failing or close to failing. For a more detailed overview, please go to **Chapter 12: Mode 6.**

# OBD Changes

### Mode $07 - Request On-Board Monitoring Test Results for Continuously Monitored Systems

Okay, you just completed a vehicle repair; but did you fix the problem? Will the MIL stay off now? Will the vehicle pass a scan tool emissions test tomorrow? The original idea behind Mode 7 is that it should test the repair for you the next time the monitor runs. If the monitor runs and fails, you should see a new **pending** DTC.

A pending DTC won't turn on the MIL; it does so only if the fault that stores the pending DTC is detected again during another monitor test, storing a DTC. The PCM says, "Fool me once, I store a pending DTC, fool me twice and I store a DTC and turn on the MIL!" Sure sounds like a good idea, but support for pending codes out in the real vehicle fleet has been uneven, especially on older OBD vehicles. In an effort to deliver the diagnostic goods, new standards say that all pending codes must be reported, starting in M/Y 2005.

### Mode $08 - Request Control of On-Board System, Test, or Component

Mode 8 is allows use of a scan tool to control an onboard system, While this is a generic mode, like Mode 6, it is implemented by a vehicle specific scan tool command. Many vehicles and scan tools do not support this feature.

### Mode $09 - Request Vehicle Information

This mode allows a technician or emission test center to retrieve the vehicle VIN, module calibration number (CALID), and Calibration Verification Number (CVN). CALIDs verify that the correct software is installed. CVNs ensure that the module software has not been altered.

By 2005 M/Y, CARB requires standard counters that show how often monitors run during actual driving, compared to CARB-defined drive cycles. The counters provide data about catalyst, $O_2$ sensor, EGR, AIR, and EVAP monitored systems.

### Mode $0A - Permanent DTCs

Permanent DTCs are MIL-illuminating codes that cannot be erased by a scan tool or by disconnecting the battery. Only the PCM can erase a Permanent DTC after running tests and determining that the fault has been repaired. Permanent DTCs are designed to close loopholes in the scan tool emissions test that allowed some vehicles to pass with a serious emissions problem. Permanent DTCs started showing up on 2009 models. For model years 2013 and later, all permanent DTCs must be available through the generic OBD interface.

At birth, OBD II was a brave adventure into uncharted territory. Many of the new software monitors used in 1996 had never been used before. To their credit, automotive engineers did a remarkable job of building this ambitious system, some of it without precedent.

Well into its second decade, OBD II is still changing: revised, enhanced, and rising to the challenge of new technology.

This chart shows its biggest changes. Expect more!

| Change | Period |
|---|---|
| Misfire monitoring under expanded operating conditions | 1997-1999 |
| Improved Catalyst Efficiency Thresholds | 1998-2002 |
| Thermostat monitoring | 2000-2002 |
| Storage of software calibration identification number | 2000-2002 |
| Calculation and storage of calibration verification number | 2000-2002 |
| 0.020 inch evaporative system leak detection | 2000-2003 |
| Positive crankcase ventilation monitoring | 2002-2004 |
| CAN Protocol Replaces Earlier OBD Protocols | 2003-2008 |
| Minimum in-use monitoring frequency requirements | 2005-2007 |
| NOx malfunction criteria for catalyst monitoring | 2005-2009 |
| Monitoring cold-start emission reduction strategies | 2006-2008 |
| Post catalyst oxygen sensor monitoring improvements | 1999-2011 |
| Primary oxygen sensor monitoring improvements | 2010-2012 |
| Permanent fault code storage protocol | 2010-2012 |
| Monitoring for air/fuel ratio imbalances between cylinders | 2011-2014 |
|  |  |

# INDEX

# Index

The Auto Emissions Bible

# Index

# Index

The Auto Emissions Bible

# Index

# Index

# OBD II J Documents

No book on OBD II would be complete without a short list of the official papers written to describe and define OBD II functions and components.

Many of these documents have been revised and enlarged over the years. The list below is hardly comprehensive, but it does list many important documents of interest to those of us performing vehicle repairs.

- **J1930**
  Terms, definitions, abbreviations, and acronyms.

- **J1962** -
  Defines the physical characteristics of the Data Link Connector.

- **J1850**
  Defines serial data protocols.

- **J1978**
  Defines the standards for the OBD-II generic scan tool

- **J1979**
  Defines diagnostic test modes.

- **J2012**
  Trouble code numbers and descriptors.

- **J2201**
  Defines the vehicle communication interface.

- **J2284**
  Defines high speed CAN

- **J2534**
  Establishes standards for a pass-through interface allowing reprogramming of the powertrain control module using a standardized, generic interface.

For more information or to order additional
copies of this book, visit us online at

*www.autorepairsolutions.com*

www.ingramcontent.com/pod-product-compliance
Lightning Source LLC
Chambersburg PA
CBHW051443170526
45166CB00001B/96